The Aha! Moment

Creature Jones in His Telly
This domestic cat invented a joke and had an idea. A creative animal!

The Aha! Moment

A Scientist's Take on Creativity

David Jones

The Johns Hopkins University Press
Baltimore

© 2012 The Johns Hopkins University Press
All rights reserved. Published 2012
Printed in the United States of America on acid-free paper
9 8 7 6 5 4 3 2 1

The Johns Hopkins University Press
2715 North Charles Street
Baltimore, Maryland 21218-4363
www.press.jhu.edu

Library of Congress Cataloging-in-Publication Data
Jones, David E. H.
The aha! moment : a scientist's take on creativity / David Jones.
    p. cm.
Includes bibliographical references and index.
ISBN-13: 978-1-4214-0330-4 (hardcover : alk. paper)
ISBN-10: 1-4214-0330-7 (hardcover : alk. paper)
ISBN-13: 978-1-4214-0331-1 (pbk. : alk. paper)
ISBN-10: 1-4214-0331-5 (pbk. : alk. paper)
1. Creative ability in science. 2. Creativity—Miscellanea. 3. Scientists—Psychology. I. Title.
Q172.5.C74J66 2011
509.2—dc22   2011010029

A catalog record for this book is available from the British Library.

Special discounts are available for bulk purchases of this book. For more information,
please contact Special Sales at 410-516-6936 or specialsales@press.jhu.edu.

The Johns Hopkins University Press uses environmentally friendly book materials,
including recycled text paper that is composed of at least 30 percent post-consumer
waste, whenever possible.

"Tell me where is fancy bred, Or in the heart or in the head? How begot, how nourishèd? Reply, reply!"

—William Shakespeare, *The Merchant of Venice*, act 3, scene 2, line 63

Contents

Preface Creativity in My Career

Having ideas! This book is a report from the front. I am a scientist, and I tell many scientific stories; but my notions of creativity include practitioners of the arts—writers, poets, composers, and other celebrated creators. I was also the crazy scientist Daedalus in *New Scientist*, and then in *Nature* and in the *Guardian* newspaper. An ideal Daedalus column started with something everyone knew and finished with something nobody could believe. Where had the argument gone wrong? Daedalus became one of the longest-running jokes in science—I wrote nearly nineteen hundred weekly columns.

In parallel with this crazy output, I did proper scientific research. My publications include serious scientific papers, as well as two books expanding and illustrating Daedalian schemes. Some of these actually came true; indeed, you cannot judge in advance whether a new idea will work out, though few of them do. Thus one Daedalian idea won a Nobel Prize for the people who finally made it work, and another was incorporated into President Ronald Reagan's proposed Star Wars project, which was a factor in ending the Cold War.

Another career I got into was making objects and experiments for TV and for science museums. Together with the Daedalus column, this steady novel practicality made me ceaselessly creative. I evolved a theory of creativity, based on my own challenges and successes. There may be other ways, but this is mine. I expound on it in chapters 1 to 4. Chapters 5 to 12 give examples of my public projects, some of the problems I encountered, and some of the feelings I had while trying out my experiments.

Creativity can often surprise its owner. At its best, a wild aha! moment suddenly gives you a new idea. I reckon it comes from a creative part of the unconscious mind, which I call the Random-Ideas Generator, or RIG (I think of it by its initials, because creativity certainly cannot be rigged, as this book will show!). Jokes and new ideas seem to use the same area of the mind; so my Daedalian jokiness—which also flows in this book—helped my creativity. You can't make contact with the RIG, or at least I never made contact with mine. And, most of its ideas are wrong. All creative people have to live with lots of failure. Worse, coming up with an idea is only a tiny part of the whole creative process. It may take years of hard work to get an RIG idea into practice.

There's a special feel to being creative. Creativity is the essential cutting edge. But ultimately, your work has to form a product of some kind. For a writer or an artist, the result has to be printed, exhibited, or otherwise put before a public. A museum curator or TV producer knows that his ideas must go in front of an audience of one sort or another. And a research scientist knows that his results will appear as a scientific paper in an academic journal. Scientific papers are detailed, formal—and boring. In chapter 5, I describe some of mine—and reveal the exciting emotions that always drive research, though papers never hint at them.

The last section of the book looks around a bit. Chapter 13 discusses some of my private creative projects, and chapter 14 tells of my life-long accumulation of facts and notions, which I now feel aided my creativity. That chapter also spells out my fascination with literary styles. Chapter 15 is a challenge to creative inventors: it recounts some inventions we need. Chapter 16 airs some of my current (quite possibly silly) questions—always a valuable stimulus to creativity. Chapter 17 condenses some of my advice on being creative.

THAT MIXED-UP CAREER OF MINE, part media freak and part serious scientist, has sparked this book. Daedalus might be deliberately silly, but my serious science often failed too. And my wild media-freakery often helped my serious science, prompting, for example, my discovery of arsenic in Napoleon's wallpaper (see chapter 16) and my studies of chemistry in space (chapter 5).

I tell lots of stories. They are not in any textbook; indeed, I dispute many textbook claims. Daedalus has leaked into many of the stories, as he also leaked into real life. I often stick my neck out and risk its being chopped off.

This strange career started in my youth. While other boys were doing sensible things like playing football and chasing girls, I built rockets and steam engines and drew animated cartoon strips and played with amateur chemistry. Much of that time I was at Eltham College, near London. A fellow pupil—David Andrews, who was much more creative than I was—became a friend. We filled notebooks with crazy drawings and ideas and generated lots of drama. Typically, he'd invent a new object and then I would modify it. Thus he built the first rockets and drew the first of our animations. He also invented the humanoid paraboloidal creatures (which we called "outfits") of several drawings in this book. Some technologies, like that of the tissue-paper fire balloon, were essentially his. Others, like photography, we played with and developed more or less separately. Still others, like electronics, I mucked about with myself. David Andrews may have been more creative than I was, but I was perhaps better at getting ideas into the world. (See chapter 2!)

My poor parents showed great heroism. They put up with my highly deviant and often destructive behavior. So did the neighbors, who often had to respond to pleas of "can I have my rocket back?" All my projects ran in parallel with the complex science curriculum of Eltham College. I went to Imperial College in London, and David to (the then) Woolwich Polytechnic. We both got bachelor's degrees in chemistry and stayed on in our institutions to get Ph.D.s, also in chemistry. Later I did postdoctoral chemical research at Imperial College.

Daedalus was born from a chance meeting with Edward Wheeler (chapter 2). Edward had studied physics at Imperial College with me, and I wrote much of the college magazine with him. One key editor was the famous Nazi sympathizer David Irving (he had those leanings even back then).

After a year of teaching at the University of Strathclyde in Glasgow, I joined the Imperial Chemical Industries Corporate Laboratory in Runcorn in northwestern England. They probably accepted me because as Daedalus of *New Scientist*, I published a crazy idea every week. (I made a special publishing deal with their patents people.) None of my industrial schemes were actualized; though I developed my theory of bicycle stability at Imperial Chemistry Industries (ICI, chapter 5).

In 1973 I left ICI and went to the University of Newcastle upon Tyne as a research fellow in the chemistry department. While there I attracted the attention of Yorkshire Television Ltd., or YTV. Soon I became their chief physical science consultant. I started to make things for them to put in front of their national audience and their cameras.

Scientific television was a shock. I had to simplify things, to leave out subtleties and evidence. The YTV science show, *Don't Ask Me* invited viewers to send questions to the so-called expert presenters. The physical science expert was Magnus Pyke—YTV loved his expansive personal style! I became the brains behind Magnus. I built the things he showed and told him what to say about them. Sadly, most public questions were televisually useless. No weekly program can survive on "Why do the wagon wheels go backward in old films?" "Where does space end?" and "Why is my reflection upside-down in the bowl of a spoon?" So I often invented the question, and we palmed it onto a member of the studio audience to ask. Of course it fitted the demonstration I had built for it.

Later I became a presenter myself, both for the BBC and on the Westdeutscher Rundfunk (WDR) German television science program *Kopf um Kopf* (*Head to Head*) based in Cologne. The German TV audience probably liked my bad German. Meanwhile, I kept on with my chemical research at Newcastle University. It helped both my science and my media-freakery. Many scientific popularizers "go native" and forget scientific detail. Not me! Indeed, when my research produced serious chemical results, I published them as from the university. I got into sober academic journals as well as popular visual and verbal media.

My brother, Peter, who by chance came to Newcastle University later and in a more formal capacity, had three children. I tried many odd scientific tricks on them (see, for example, chapters 11 and 12). We had a lot of fun, but I also watched for TV appeal. The new Joneses learned a lot of strange science!

This book expounds on many other creative matters—whether or not I made sense of them, or got anywhere with them. You may have seen some of my TV shows and may disagree with my arguments. But read on . . .

The Aha! Moment

1

A Theory of Creativity

There are two ways of solving a problem. If a rational solution exists, you apply it. This just takes whatever willpower is needed to bash out the right long multiplication or to construct and solve the correct equation or whatever. But suppose there is no rational solution? Then to solve the problem you have to be creative. You need a new idea.

Take, for example, the problem of remodeling the kitchen. Many of us have faced this task at least once. The basic problem is to make the best use of your space. I suppose you could tackle it purely rationally. First you would define some complex evaluation function giving the utility of the kitchen as a function of the position of the fridge, stove, dishwasher, cupboards, table, and so on. This would give you some nightmarish equation in many dimensions, which you would differentiate to obtain the maxima of the corresponding hypersurface. The largest maximum, when you had it, would represent the best possible arrangement. It would be a formidable mathematical exercise.

But nobody would tackle the problem like that. The normal human approach would use ideas. "Suppose we put the stove in that corner. This means that the dishwasher has to go over there. The fridge can fit in this space next to cupboard number 1, and cupboard number 2 can go next to the sink. Ah, but then you can't open cupboard number 1 because the fridge blocks it. Hmm. Well, how about putting the fridge where the dishwasher is now and moving cupboard number 2 next to the stove?" You imagine possible solutions and work out their consequences. Sooner or later, one of these ideas turns out to be satisfactory or so close to satisfactory that a simple rational modification will complete the solution. Most ideas fail in practice, so everyone trying to be creative has to live with lots of failures. It doesn't matter: you discard the ones that don't work.

Precisely the same style of thinking applies in science and technology. You cannot, in logic, deduce a theory from the data it must explain or a machine from the need it must fulfill. So a scientist or technologist dreams up possible theories or possible machines and sees whether they fit. Most of the time they don't. Sometimes you have to devise an experiment, or even a whole program of them, to clarify the problem. I have wasted vast amounts of time asking the wrong question or building an apparatus that merely shuts off one stupid area of inquiry. But even with hindsight I cannot advise any other way to go.

Linus Pauling, who won his first Nobel Prize in 1954 for chemistry, was once asked how he came by his notions. He said that "he had a lot of ideas, and threw the bad ones away." His reply supports my fear that most ideas are bad. Sir Peter Medawar, who was a Nobel winner in 1960 for medicine and physiology, was more precise: "for all the use it has been to science, about 80% of my time has been wasted." I reckon 80% wasted is very good. It makes me compare the many experiments I have done with the few that I have actually described and published. I can tighten the statistics even further. At Newcastle University, I once spent about a year of my life building an apparatus to create a chemical "garden" in space. All that time I was trying to predict whether it was feasible and what would happen. I was being a serious scientist indeed. Mercifully it worked, and I found out (see chapter 5).[1]

It did two things I had predicted. It failed to do two things I had predicted. It did six things I did not predict. My score was thus 2 out of 10, or 80% wrong, just the proportion that Peter Medawar would have expected. When I worked in industrial research (at ICI Ltd.), we had a rule of thumb that was even worse. Of the ideas suggested by the research department, 10% might make it to the pilot stage. Of the pilot schemes, 10% might make it to production. Of the production processes, 10% might make big money for the company.

So serious professional scientific ideas fail 80 to 90% of the time. Weirdly, so does sheer frivolity. One of my major creative activities was the weekly Daedalus column, which I wrote first for *New Scientist* and later for *Nature* and the *Guardian* newspaper. It had to be scientifically funny. I was free, indeed obliged, to put forward great scientific absurdities. But despite my best endeavors, these mad Daedalian schemes kept coming true on me. About 20% of them made some sort of contact with reality. One earned a Nobel Prize for the people who finally made it work. Another was the first suggestion of a scheme later turned into reality by the United States Air Force.

Another part of my creative activities was building things to show on television or to exhibit in science museums. I might suddenly have an idea, followed by a struggle to make it practical. Often I dug through a lot of possibilities first. The producer or manager of the project filtered my initial suggestions rapidly. Perhaps 20% of them survived. In the end, the audience only saw the one finished, working product. Yet sometimes one of those losers felt to me as if it had potential. I could only make a note of it and wait for the chance to develop it in the future. This brings up the style of creativity I have called "feminine" (see chapter 2), which means that you have a lot of ideas over time, not even vaguely aimed at the same goal. As the feasible ideas emerge, you apply them to some long-term oeuvre that you are working on.

I came to depend on my output of ideas, unfeasible though most of them were. I grew to respect their unconscious source. But whatever that unconscious source was, it knew very little science. Absurdities did not bother it. I just let them come and became much more tolerant of silly notions than most scientists.

How do you get ideas? Nobody can have a truly new idea—all we can do is to combine existing facts or notions, gained by observations or the remarks of others. In this book, I shall argue that you need a vast subconscious mass of remembered data. Thus the kitchen remodeling example assumes that the problem solver has worked in a kitchen, has talked with others who have worked in a kitchen, and has accumulated a wide variety of kitchen experiences good and bad.

Mental Structure

Human beings have developed from animals. We have not evolved anything new, but we have greatly expanded many animal abilities. Thus many animals also have an unconscious mind. According to Robert Trivers,[2] the animal unconscious exists to hold "personal political information" safe. This will be its private feelings about the animals close to it and

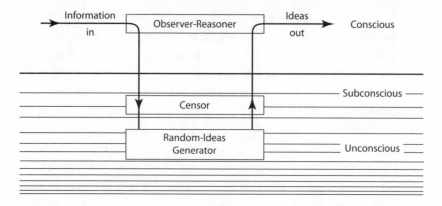

FIGURE 1.1   Human Mental Structure
In my model, the Observer-Reasoner is in the conscious mind, and we are aware of it. The Censor is in the subconscious mind, and we know little about it. The Random Ideas-Generator, or RIG, is in the unconscious mind. We cannot contact it but depend on it to have ideas and pass them upward.

the personal pressures it feels. It can then lie about all this to the other animals. The best liars consciously believe their lies. So, to deceive the other animals, an animal has to deceive itself about, for example, its position in the pecking order, the characters of other animals in the pack, its special friends or enemies or teachers or pupils in that pack, or its hopes of becoming pack leader.

Our human mental structure resembles that of pack animals, though modern psychology knows more about it. My simple subjective model of the human mind (fig. 1.1) includes much of what we can feel from inside. I call the top element the Observer-Reasoner. This is the part of the mind we are conscious of. It studies incoming data from the senses, looks critically at ideas that come "upstairs" from below, and plans our next actions. Below it is the subconscious, which is the bridge between the conscious and unconscious minds. Skills that we mastered long ago, such as how to balance on two feet, swim, use a hammer, or ride a bicycle, are stored in the subconscious. We can access these rapidly as needed. Crucially, our linguistic skills are subconscious. When we are talking or writing, or listening to the speech of others, we are accessing our vocabulary with lightning speed. Our verbal skills depend entirely on our fast retrieval of words and their meanings from the subconscious.

Further down in the model is the true unconscious mind. It holds the material that we can almost never get at. As with any pack animal, this is mainly "personal political information"; as with them, what we can get at seems deceptive and evasive stuff. It may reveal itself, in a disguised sort of way, in dreams. Psychiatrists have claimed that recalled human dreams sometimes usefully reveal some aspect of personal politics. Freud even wrote a book on the interpretation of dreams. The manifest content of a dream is what it seems to be about; behind that is the latent content, its true personal meaning. Freud's book encouraged a huge industry: that of interpreting dreams. I reckon that the human unconscious mind generates not only dreams, but jokes and creative ideas too (via the entity I call the Random-Ideas Generator).

The Random-Ideas Generator, or RIG

Most of the unconscious mind deals with personal politics. The creative part is (I feel) only a small offshoot. In my model I call it the Random-Ideas Generator, or RIG. (In my mind, it has three letters, R-I-G, and is not to be confused with a rig of any sort.) I feel it has access to all the information stored in the subconscious and the unconscious minds, which in my diagram shade into one another. The RIG combines the things you have stored and sometimes pushes some combination or generalization "upstairs." I don't know how it works, or what sort of things it tries to combine. In my ignorance I just call it "random." But it probably has a range of facts it knows and likes and can play with and some awareness of the problems that are currently bothering the Observer-Reasoner. In the rest of this book, I often refer to my own RIG. I am guessing. I may be talking about any part of my deep mind that rarely communicates with conscious awareness.

The whole set of our mental entities, conscious and unconscious, form one unit, our "self." This experience seems usual; but the mathematician Michael Alder might disagree. In 2004 he was writing in the computer language LaTeX when he had the sense of being "taken over" by something inside him. It bashed out a piece of mathematics through his fingers and onto his screen and paper. He had this "intellectual diarrhea," as he put it, for about a week. His internal entity was often a bit sloppy, and what it wrote sometimes needed to be tightened up. But apart from that tightening, Mike had no sense of "ownership" of the product. My guess is that his RIG had generated the work seemingly as a separate personality. But most people seem to "own" their whole mental structure. Thus the poet Rainer Maria Rilke had much the same experience as Alder (see chapter 3), accepted the product as his own, and called the experience "utterance and release."

When the RIG has pushed up an idea, the Observer-Reasoner checks whether it will work (usually it won't). The RIG is active all the time, and sometimes pushes a creative notion upstairs quite spontaneously—that aha! moment when we get a new idea. In my guise as Daedalus, I also suggested that, in women, the unconscious may even influence aspects of reproductive strategy (see chapter 4). Even animals may have an RIG (see the story about the cat and the bathtub, below).

The Censor

Opposing the RIG—indeed opposing the whole unconscious mind—is the Censor (see fig. 1.1). It has to keep the Observer-Reasoner safely apart from the "lower" regions. So it is aware of all facets of the mind from the Observer-Reasoner to the RIG and everything in between. One way of boosting your creativity is to modify its censorious strategy (see chapter 2).

The Censor has a tricky job, in fact several tricky jobs. First, it has to prevent personal political information from surfacing. When such information has to come up, as perhaps it has to in a dream, the Censor only lets it through when it is distorted enough to make no sense. It faces a second tricky job when notions from the RIG want to make their way to the conscious mind. Most of these notions are "duds." Even the survivors, which the Censor allows upstairs, are at least 80% duds. If the Censor is too restrictive, it hampers creativity; if it is too permissive, it bothers the Observer-Reasoner with a lot of nonsense.

A third Censorious job is to prevent nonsense, untruths, unimportant observations, or worrying heresies from getting downstairs for the unconscious mind to play with. For example, many loyal Nazis contrived not to hear about the concentration camps or not to remember about them. Only an effective Censor saved their worldview from trouble!

I am not attacking the Censor as such. Its protective function matters, and that is why we have it. Down in the unconscious mind, all sorts of dangerous absurdities and mad possibilities are being tossed around. This is where Freud's Id lives; in a way the RIG is the intellectual wing of the Id. The RIG is valuable in play (see chapter 4), imagination, and other childish pursuits. Indeed, in human life, play and the RIG develop much earlier than reason does. In childish play a chair can be a galleon, a dragon, or anything else. Later in life, reason comes down, as Victorian respectability must have come down on the libertine Regency world.

Ted Hughes (a former poet laureate) has even personalized his Censor. He has called it the "inner police system of the writer." He suspects (as I do) that creativity consists at least in part of outwitting the Censor. Wendy Cope has bewailed the plight of the Censor in a splendid poem that parodies Sir William S. Gilbert's famous song "A Policeman's Lot."[3] Her parody imagines a Censor as a police figure "patrolling the unconscious of Ted Hughes."

More about the RIG

The Random-Ideas Generator is seldom overawed by the conscious, repressive, rational mind above it. I imagine it playing around with the ideas and observations that get "downstairs" to it. It cannot think, only imagine; and its combinations seem dominated by aesthetic feelings such as beauty and range. The mathematician Henri Poincaré has commented on sudden insights which, after study, have turned out wrong. He has noted that such an insight, had it been correct, would have been very elegant.

So it is not surprising that most sudden RIG notions are wrong. The RIG may make many odd irrelevant combinations of ideas that it does not present to the Censor. It knows they will be rejected! And it seems not to care about scientific facts or laws. Mine has (perhaps) grasped the law of conservation of energy and tends not to imagine perpetual-motion machines (see chapter 14). But it ignores many other physical laws.

Yet we all depend on this quirky mental object to solve practical problems. With a fairly simple problem (such as arranging the kitchen appliances), the RIG has ideas almost on demand and passes them up as rapidly as the Observer-Reasoner can evaluate them. More complex projects, like a symphony or a novel or a scientific theory, can occupy it for years. Often its insights come up at intervals as single "fractional" contributions (see chapter 3).

You cannot have an intellectual relationship with your unconscious mind. Like a pet animal, it's an emotional entity. But you can be fond of it and pleased with what it gives you. It can then go where you can't go and can bring you things you cannot get (and indeed, may not want). Or you can be dissatisfied with it, when it may clam up or claim not to be there. It may even hide things from you.

Dreams

We all dream several times a night. Head electrodes can look for signs of "dreamy" brain activity. Electrodes around the eyes, or the muscles that drive them, can look for signs of eye movements under the closed lids. This "rapid eye movement," or REM, sleep can be seen in human beings and even in many animals. Patricia Garfield has written much on dreaming and says that all mammals dream, except the spiny anteater.[4] Perhaps it is hard to attach electrodes to a spiny anteater.

There is no good theory of dreams. One theory notes that we all need to sleep, that we dream regularly, and that we forget our dreams very rapidly. A dream may be a way of discarding much of the day's memories. Any important new stuff is added to the brain's long-term storage; yet we only have one brain to hold a growing lifetime of recollection. Pure trivia (such as innumerable breakfast menus) must be pruned ruthlessly and often. Dreams show the mechanism at work. Sadly, they make little sense as a daily diary of rejected trivia. Daedalus has claimed that they are in the brain's internal "machine code," not the high-level language in which we consciously think (see chapter 4). So perhaps they are trivia after all but in machine code.

Freud and later analysts of the dream have a deeper interpretation. They reckon that a dream represents our current personal struggles in camouflaged form. To provide hard data for any theory, Daedalus has suggested (chapter 4) a way of recording dreams. My own guess is that dreams are a random scan of the unconscious mind, disguised or modified to get past the Censor. They are mainly derived from personal political information, but a few of them may contain jokes or technical ideas (which I call "technical dreams"). I have only had one useful technical dream in my life (see chapter 11), but some commentators have enthused about them.[5] Here are some that I have noted.

ELIAS HOWE AND THE SPEAR

The most interesting technical dream was perhaps that of Elias Howe, one of the inventors of the sewing machine. He had struggled with the idea for years. He started by trying to imitate his wife's hand as she sewed. One night he dreamed that he was in the grip of a savage king, who had given him 24 hours to build a sewing machine and to make it sew, or die. In his dream he saw himself defeated and led out to execution by warriors with spears—spears that had a hole in the blade, near the point (fig. 1.2). He woke up. That was it! The eye was in the wrong end of the needle! It should be at the pointed end!

It took Elias Howe many years to build his machine, patent it, and enforce his patents against Singer (who had infringed them), but all modern sewing machines have a needle with the eye near the point.

LOEWI AND THE FROGS

Another technical dream was that of Otto Loewi. He was a physiologist, and the idea of the chemical neurotransmitter (for passing a chemical signal from one nerve to another) came to him in a dream of 1920. He woke up and made a note but found it cryptic and unreadable in the morning. He simply hoped the dream would come again. It did; this time he made a more careful note. Later, he tested his revelation. He tried an experiment in his laboratory, on two frogs' hearts, dissected out but beating. He could slow one down by applying a liquid from the other—thus proving that a chemical conveyed the information. At the time he called the chemical "Vagusstoff," but we now know it to be acetylcholine. It rapidly breaks down in the body, but luckily Loewi used a species of frog in which it lasts long enough. Later he found that he had first had the idea in 1903 but had discarded it with many other ideas. He suggested that his unconscious mind remembered that notion and re-presented it to him in his dream. Anyway, he built on that insight and won the Nobel Prize for medicine in 1936. Today chemical neurotransmitters are central to all nerve theory.

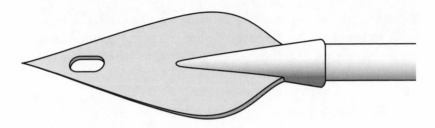

FIGURE 1.2   My Guess at Elias Howe's Dream
It helped him invent the sewing machine.

BOYS AND THE GAS METER

A third technical dream was that of Charles Boys, the mighty experimental physicist (chapter 11). Among many other things, he was a gas examiner, a scientist who ensures that a gas-meter records correctly, no matter what the pressure or temperature of the gas it is handling. His Guthrie lecture of 1934 details how he had a dream describing a greatly improved gas meter. He got up, went to his laboratory, and blew the glass bulb for a prototype meter at 6 a.m.

ARTISTS ALSO DREAM

The composer Giuseppe Tartini created the famous "Devil's Trill" in a dream. In it he gave the Devil a violin. The Devil played an amazing composition, which Tartini tried imperfectly to jot down on waking.

Robert Louis Stevenson wrote his celebrated story *Dr Jekyll and Mr Hyde* as a result of several dreams, during one of which his sleeping cries of horror caused his wife to wake him. His unconscious mind did so much work on that story that he completed the first draft in only three days.

Paul McCartney was dreaming one night and woke with the tune of "Yesterday" in his head. He thought he had heard it somewhere. After a month of puzzlement, he decided that he had invented it. So he wrote words for it.

Daydreams

In this account of creative dreams I have included neither Friedrich Kekulé's notions of chemical structure nor his cyclic structure for benzene. Did he actually dream them? Chemical structure came to him while he was on an omnibus in the London district of Clapham. Later he saw the circular benzene molecule as he was gazing dreamily into his fire. I suspect that he was half-awake, in that "dozy" state that I regard as more creative than full sleep. Dmitri Mendeleev may have conceived his chemical periodic table in that state—nodding dozily after playing with a set of cards he had marked with the names and properties of the chemical elements.

The state of reverie or daydreaming, maybe lying down but not actually asleep, can be very creative. I applaud the experience of Kekulé and Mendeleev above. Hideki Yukawa, perhaps, had the idea of the nuclear meson in that reverie state; and the electromagnetist Eric Laithwaite has saluted it, though he acknowledges that mainly it produces rubbish. In it Richard Wilhelm Wagner imagined the overture of "Das Rheingold," while John Fowles created many characters and plots for novels. Does it bypass the Censor and let you into the RIG directly?

Jokes

Humor, like creativity, probably resides in the unconscious mind. The physicist John Wheeler has claimed that creative physical theory gains from a sort of playfulness. He extolled "this bounce, which I always associate with fun in science, kicking things around. It's not quite joking, but it has some of the lightness of joking. It's exploring ideas." Arthur Koestler has gone further.[6] He identified jokiness and creativity, which in his view are both the result of a collision between two incompatible visions of the world. I think he is on to something. Humor and creativity do indeed use the same sort of mental ability. One of his examples is a Frenchman who finds his wife in the arms of a bishop. He goes to the window and starts blessing the people in the street. "What are you doing?" cries the distraught wife. "Monsignor is performing my functions, so I am performing his."[7] Peter Medawar has denounced Koestler's claims, on the grounds that a failed scientific theory ought to appear in retrospect as a huge joke. The slanging match between them is reproduced in Medawar's *Pluto's Republic*.[8] I can see both sides of the argument. In my view, a failed theory (a line that does not fit a set of points, say) is not a change radical enough to qualify as a joke, which completely changes the context set up by its beginning. My sense is that jokes and new theories are different, but both in some way go along the same mental axis. Both are in the province of the unconscious mind; both depend on interpreting matters in a novel light.

Indeed, the best jokes make a sort of initial sense; they sink a little way before they explode. My favorite example is Groucho Marx's joke: "I do not care to belong to any club that would have me as a member." In 2001 there was a competition for the funniest joke ever, held (of course) on the Internet. My entry would have been a joke of 1845, published in Punch: "Advice to persons about to marry. Don't." (Marriage advice was common at the time; it usually advised the couple to buy something.) Another potential entry occurred in the film *Casablanca* of 1942. A French policeman says, "Round up the usual suspects." Both jokes are still funny; both have entered the language; both completely change the context in which they were set up.

Of course, we humans have taken jokes very far compared with animals, who seem to live highly serious lives. But the domestic cat Creature Jones (seen in the frontispiece) may have invented a joke.

A Cat's Joke—Creature Jones and the Side Door

When I was growing up, my family was a very jokey one. We played jokes on each other and on the family cat. My guess is that Creature Jones somehow picked up this aspect of family culture. Indeed, he may have invented a joke himself—or at least this is my interpretation. We had a long back yard, maybe 60 meters long and 15 meters wide. At the far end there was a scruffy bit of land. The compost heap went there, we played crude ballgames there, I tried various scientific experiments there, Mum put the rougher plants there. So there was often a human being at the bottom of the yard. Creature Jones, who liked human company, would often be there too. Then the human being would walk up from the yard toward the house, perhaps to get some implement or to start another domestic task. The cat would sit at the bottom of the yard as if he intended to sit there all day. When he judged the time was right, he would run like mad to the side door, going much faster than the walking human and would then sit there, as if he had been waiting all day to be let in. The human might be puzzled. "I could have sworn I left that cat at the bottom of the yard! But here he is at the side door." My impression was that Creature Jones was very pleased with himself after this performance. When I was around I would tell him what a wonderful cat he was, what a splendid joke he had played, and how much we had all enjoyed it.

Even the rare human attribute of "genius" resides, in my view, in the unconscious mind. Thomas Edison said that genius was 1% inspiration and 99% perspiration, which in science and technology may be about the right balance. Even so, that inspirational 1% is crucial. And it may totally baffle your colleagues. Thus Gregg Herken said of the great physicist Andrei Sakharov, that, "even after we understand what he has done, the process by which he has done it is completely dark."

The Importance of Play

Play is the division of the world into a "real" world and a "pretend" world. Many young animals demonstrate that division. In kittens or puppies, say, play consists almost entirely of mock fights—doing in fun what they will later do for real as adults. They know very well that this is play and do not damage each other. Neoteny (the extension of childish behavior into the adult) marks the domestication of an animal. Pet owners, and even farmers, prefer an animal to have a certain youthful flexibility. The domestic cat Creature Jones would sometimes "challenge" me to a fight. I would get hold of a bamboo cane and poke it along the ground at him, letting him pounce on it and worry it as if it were a snake. After 10 minutes or so he would lose interest. Both of us knew that this was only "play." By contrast, if he was chasing a mouse that ran into a hole he would watch that hole for hours—this was "real" and not play.

In the same sort of way, the unconscious mind does not think about information. It is an emotional entity and plays with it. Yet every scientific theory starts as play: let's pretend that this obeys that law, or whatever. What follows? Mostly, the thing is clearly not true. Occasionally, it makes enough sense to get exciting. But the beginning is always a "silly idea," a pretence, play. "Let's pretend that this substance has this molecular structure," says the chemist. And maybe it does. The pretence may fit so well that the chemist "buys" that structure and regards the problem he is working on as solved. It becomes part of his known or relatively unquestioned world. Or he may nag and worry about some detail that does not fit (see chapter 5).

Curiosity

Curiosity is being interested in experience, noticing it, and remembering it. It is an important part of being creative. You have to spot, and store downstairs, a huge fund of information and experience. Much of it is stored in the subconscious mind. In my mental model, both the Observer-Reasoner and the RIG can get at it. So it is available for the RIG to play with. Occasionally the RIG combines a set of notions that get down to it and sends a promising combination up as a new idea. You become aware of it, and the Observer-Reasoner looks at it. It also looks at information from the outside world (see fig. 1.1). So it pays to develop your curiosity. Creatives have been described as "noticers" and as having a "low pain threshold"—things bother them. By contrast, animals keep their inherent curiosity firmly in check; they cannot easily afford its risks. When the notorious tsunami struck the shores of the Indian Ocean on 26 December 2004, many animals detected its soil infrasonics and fled to high ground where they were relatively safe. Many humans, even those who suspected what might be coming, were curious; they went to look and were killed. Yet curiosity is a powerful human motivation, and I approve of it. "Curiosity killed the cat," says the proverb. Maybe; but it made cats what they are. Without it, they'd be rabbits. I like to think that human curiosity is inspired by our unconscious mind, always seeking new information, and goading the Observer-Reasoner into acquiring it. I have more thoughts on human curiosity in chapter 2.

What does the Observer-Reasoner do with the observations that the senses bring to it? Expertise, of course, will always show. At a memorial, for example, a builder may notice the material of construction, a botanist may spot the species and siting of the growing lichens, a typographer will be interested in the form of the carved lettering, a historian may judge its style and character and guess its age. And if you are worrying about the kitchen appliances (say), you may notice any feature that seems relevant to your problem. But here I am advocating a more general curiosity and inquisitiveness. The Observer-Reasoner of a creative will be intrigued by, and may notice, almost anything. A small amount of what it notices may get down to the RIG, which will play with it. From the resulting rich mixture—observations noticed, data stored in memory, a remark by somebody else, a scheme pushed up by the RIG to the conscious mind via the Censor—the Observer-Reasoner may get a new idea. The next stage of the process will probably need a lot of hard conscious work. You have to evaluate your new idea: bear it in mind, worry about it, and work on it. Isaac Newton was once asked how he made his discoveries. He replied, "By always thinking unto them. I . . . wait till the first dawnings open little by little into the full light." To him, as to any creative, steady evaluation mattered. You may conclude that your new idea is just wrong—after all, about 80% of them are. Or you may imagine some simple twiddle that makes it seem true and important. Even if it is nonsensical as a whole, it may have some element which seems somehow "promising." It pays to record or remember such a "fractional idea." Even so, you may have to wait for months or years to come across something that matches, completes, or adds to it. This has happened to me. Isaac Newton held his gravitational theory back for years—maybe until a new measurement of the Earth's diameter made his calculations fit. To get further, your new idea will probably need a lot of rational examination and intense evaluation. You may have trouble just explaining it to other people! (I discuss this sort of effort in chapter 2.) Bringing your idea to fruition will need a lot of hard conscious work, often practical. Edison's remark that genius is 99% perspiration probably came from bitter experience.

Stories of Creativity

I have many examples of creativity. They include stories of my own, talks with originators, and writings I have looked at. They all seem to tell of events "off the job." In 1931 Washington Platt and R. A. Baker circulated a questionnaire about creative moments, "hunches," as they called them, in chemical research.[9] Many respondents recommended journeys. One consulting chemist told them that he got good ideas in the train (e.g., a Pullman berth) because he could not be interrupted and knew he could not be interrupted. Another recommended being the driver on a car journey. Such a driver is occupied by many trivial tasks; these saturate the conscious mind and keep it busy and out of the way. I have particularly noticed the absence of any creative time-scale. You may get an idea after years of work or you may get it immediately. For more of my musings about time and the RIG, see chapter 3.

A CAT'S IDEA—CREATURE JONES AND THE BATHTUB

I am sorry to say that I did not see the crucial step. And I have no idea how long Creature Jones's unconscious mind had nurtured his notion before "pushing it upstairs" into consciousness and action. The cat had its territory, as domestic animals always do. He "owned" our front and back yard, for which his human family supported his claims. Beyond the garden fences, he had to negotiate boundaries with the domestic animals of the neighbors, where they had them.

There were also wild animals around, squirrels and hedgehogs and pigeons and such, with whom no negotiation was possible, and with whom he lived in a state of constant war. But inside our house he was supreme. No wild animals or rival cats ever came in; he was with his human family. It was his domain. He knew every room in the house, upstairs and downstairs, and went into each with pride and confidence.

So it upset me that he was frightened of our bathtub. He rather liked the bathroom itself; he would go into it and weigh himself on the floor scales and sometimes he would sit on the bathroom stool. He was a trusting cat. If I picked him up, he would nestle in my arms and be happy to go where his human friend wanted to take him. But if I took him near the tub, he would bite and scratch and try to get down and behave most uncharacteristically. He was seriously frightened of that bathtub. I don't think he knew that there might be water in it. I just think it made no sense to his eyes. To him, it was a great white hole in space in which he would fall for ever and ever. I once stood in the tub for him and showed him that you did not fall for ever and ever. He stared at me, but his fear did not diminish. I gave up my program of bath education for cats. I accepted reluctantly that in the very middle of his domain there was an object of which he was deeply and permanently afraid.

Then one day, in the summer of 1966, my parents went on vacation. My brother was away, I was living in London in an apartment flat. Mum arranged that the cat would stay at home and that our neighbor would feed him. On the weekend, I came down from London and checked that the house was all right. When I came down, Creature Jones was very pleased to see me. I let him into the house. Then I busied myself downstairs. Probably I looked at the mail to see if anything had arrived that needed urgent action; probably I went into the kitchen to make myself a cup of tea. Then I remembered my domestic duty and went around the house, checking that the power was off on all the appliances, the faucets weren't dripping , and so on.

I went into the bathroom—and there was Creature Jones, standing in the tub! When he saw me he paraded up and down in clear delight. I was amazed. With no obvious goal or motive, this old cat had confronted and overcome a fear that had ridden him all his life. Alone in the bathroom, Creature Jones had taken his life in his paddy paws and had leapt into the tub. I wish I had been there to see him, but perhaps he needed to be alone. Later I guessed what must have been going on in his mind. His domestic routine had been totally upset. His human family had vanished; for days he had been excluded from his domain. When I let him in he had to check every room in the house for signs of disturbance, for wild animals or rival cats. He did not find any. His keen sense of smell did not even detect much human odor, even in the bathroom. And perhaps the complete disruption of his way of life encouraged his unconscious mind to push "upstairs" a bold experiment it had been holding. He had an idea—that of leaping into the tub.

When my parents came back home, I told them of the dramatic thing their cat had done. Thereafter, Mum would occasionally hear a yowl of triumph from upstairs and would go up to find Creature Jones reasserting his victory over the tub. Once she found him in the neighbor's tub. He could beat any bathtub in the street, this cat!

BOWERS IN THE CREVASSE

Captain Robert Scott's Antarctic expedition, begun in 1910, is one of the heroic stories of British exploration. Part of it was a winter trek to obtain penguin eggs. The three men of the party, Apsley Cherry-Garrard, Edward Wilson, and Henry Bowers, endured appalling hardships on their 5-week journey. Wilson and Bowers perished during Scott's fatal polar journey of 1912, but Cherry-Garrard survived to write of that trek. Among the hazards the party faced were many ice crevasses.

Some were invisible in the dark of the Antarctic winter, and some were bridged at the top by soft snow. The men roped themselves together for safety. While leading them, Bowers fell into a crevasse. The two men on the surface could not haul him out because the loaded rope of his harness froze tight on the lip of the crevasse. Wilson shouted down, "What do you want?" In the circumstances, this was perhaps the most ridiculous question possible. But in a crucial burst of inspiration, Bowers asked for a rope with a bowline (a tied loop) on the end. He put his foot and weight in the bowline and could raise himself about 30 centimeters. Relieved of his weight, the main harness rope could then be pulled up a bit. Bowers put his weight back on the harness, and the bowline could then be pulled up a bit. And so on. Bowers was slowly ratcheted upward to safety, and this two-rope method became the standard way of getting an explorer out of a crevasse. It had been invented on the spur of the moment by a man hanging in a crevasse himself!

TWO BICYCLE IDEAS

In 1987 Adrian Spooner, head of classics at Park View School in Chester le Street near Newcastle, began to assemble a book. Its title was to be *Lingo: A Course on Words and How to Use Them*. It would teach children ages 11 to 14 how to use English by showing the derivation of words from Latin and Greek. But Adrian could not see how to set the idea up in book form.

He was an enthusiastic amateur actor and regularly went to the People's Theatre in Newcastle by bicycle. On this occasion he had gone three-quarters around the Benton traffic circle, when he saw how a tripartite division of the book would structure it just as he wanted. His bicycle veered dangerously as he had the idea, and he nearly caused an accident. He had to stop on the middle of the circle until he had returned to the real world. The final book (published in 1988 by the Bristol Classical Press) uses a classical myth to introduce some useful high-level English words derived from Latin or Greek. The second part gives the etymology of these English words, and the third part shows how to link them into sentences.

In September 1960 the Reverend Jack Rutherford became the new vicar in St Philip's church at Arthur's Hill, Newcastle. For years he had been a Tyneside curate and had long wanted a church and a parish. But St Philip's was not to his taste. He accepted it grudgingly; he was an unpopular priest, and his incumbency started badly. One autumn day in 1960 he was pedaling up Stanhope Street on his bicycle toward his vicarage. He was passing Corrigall's drugstore when, as he put it, "God hit him on the back of the head." God said, "Stop complaining. It's not your parish, it's *my* parish! Now get on with it!" Jack seems not to have risked a bicycle accident, but his attitude changed entirely. He became an enthusiastic and inspiring priest. Next summer the church was so packed that people were clustering on the pipes that ran along the aisles because the pews were all full.

(The most famous case of a divine message delivered suddenly on a journey is, perhaps, that of Saul on the Damascus Road.[10] Saul—later Saint Paul—was not on a bicycle at the time. Caravaggio's painting of the event shows him apparently having fallen from a horse.)

BLACK AND THE NEGATIVE-FEEDBACK AMPLIFIER

Negative feedback is today one of the most pervasive notions in science. In technology, it lies behind any number of control systems; in biology, it lurks in all evolution theory and all in psychology and physiology. And in electronics, negative feedback is fundamental to almost all analogue systems. And yet the negative-feedback amplifier was only invented in 1927.

Harold Black took six years of hard work to create it. The story began in 1921 in the Western Electric Company's laboratory at 463 West Street, New York City.[11] (In 1925 this became part of the Bell Telephone laboratories). Harold Black had to improve the chains of amplifiers needed for long-line, multi-channel telephone traffic. The existing amplifiers introduced such distortion that a chain of them made speech almost unintelligible.

In 1923 Black attended a lecture by the electrical genius Charles Proteus Steinmetz. It sent him right back to first principles. Soon he invented the feedforward amplifier. This greatly reduced distortion, proving that it could be done. But it was very complex and needed such frequent adjustment that it was not commercially attractive.

The negative-feedback amplifier came to Black on 2 August 1927. He was crossing the Hudson River on his way to work, on the old Lackawanna Ferry, a sort of sea-going conservatory. If she had been built for this one journey, and then sunk with the paint still wet on her, she would still have been an excellent bargain. For decades Black wondered in vain why the idea had come to him at that moment. "All I know is that after several years of hard work on the problem, I suddenly realized that if I fed the amplifier output back to the input, in reverse phase, and kept the device from oscillating, I would have exactly what I wanted." Still on the Lackawanna Ferry, he sketched a circuit and the basic equations on his copy of the *New York Times* and signed it. Twenty minutes later he reached the laboratory; his boss witnessed, understood, and signed the paper too. By establishing an exact time of invention, it became a basic document for the ensuing patent. That patent took a further nine years to complete—the scheme seemed impossible, like a perpetual-motion machine. (I discuss such machines in chapter 14.)

CHANDRASEKHAR AND THE BLACK HOLE

In 1930 Subrahmanyan Chandrasekhar was a very bright Indian student of 20. The Indian government granted him a scholarship to go and study for his Ph.D. at Cambridge University under Ralph Fowler. So on 31 July 1930 he boarded a ship in Madras, heading for Europe. In the days that followed, being a passenger on the ship with no duties, he read Fowler's paper applying quantum mechanics to white dwarf stars. This type of star is typically about 50,000 kilometers across. Chandrasekhar began to add his knowledge of relativity to the paper—this combination of principles was new to stellar theory.

He concluded that above a particular weight (now known as Chandrasekhar's limit, about 1.4 solar masses) a white dwarf star was unstable. Its gravitational attraction would ultimately overwhelm its nuclear repulsion and it would collapse to a tiny object. It might swallow its own light. These concepts would have been a great extension to the laws of physics as then understood and a mighty step in the theory of the black hole. At Cambridge, Fowler was not convinced by Chandrasekhar's calculation, but he had the support of Niels Bohr in the renowned physics school of the University of Copenhagen. In 1935 Arthur Eddington, one of the great pigheaded geniuses of cosmology, used a Royal Astronomical Society meeting to denounce Chandrasekhar's theory of stellar collapse. Chandrasekhar went to the United States and became a major physicist in his own right (he was awarded the Nobel Prize for Physics in 1983). He built his career at the University of Chicago. The calculation he started on that voyage is now a centerpiece of modern physics!

PAULING AND THE OXYGEN METER

In 1939, the U.S. military authorities organized a conference in Washington, D.C. They presented about 20 top American scientists with a list of problems they wanted solved. One of them was determining the level of oxygen in an atmosphere. The army (which commanded the air force) wanted this for aircraft, and the navy wanted it for submarines.

Linus Pauling attended this conference. He traveled from California to Washington by train, a journey of several days. On his trip back, he began to muse on the military's troubles. He reckoned he could solve the oxygen problem. Oxygen, as he well knew, has a magnetic moment, a weak molecular magnetism (unlike the other common gases). By the time his train arrived back in California, Pauling had designed a simple magnetic oxygen meter for gases. He gave the design to his students, who built the prototype in a week.

Many copies of the finished product were made for the military. And it is still in use! Beckman, the instrument company, makes it. These days the Beckman oxygen meter is intended mainly for the incubators of premature babies, who need a specific proportion of oxygen in their air. Too little may damage the baby's health generally; too much may damage its eyes.

ARLEN AND "OVER THE RAINBOW"

The MGM film *The Wizard of Oz* began to take shape in 1937. It was to be a vehicle for Judy Garland, then aged about 16. She had to sing a central song. The contract to write that song was awarded by MGM to Harold Arlen (for the music) and Yip Harburg (for the words). It was a daunting challenge. Indeed, the authors overran their contract time. So Arlen was working for nothing when he was being driven somewhere by his wife, and suddenly said, "Stop the car! Stop the car!" She did so; and Arlen jotted down the first rendering of the tune of "Over the Rainbow." Even so, big troubles lay ahead. One early version of the song was "far too operatic, wrong for a young star," as Harburg mused. And when the song was completed, two film directors did not like it. Louis Mayer was the last M in MGM. He may have been a corrupt barbarian, but he had some feeling for the film medium.

Finally he said, "Let the boys have their song." It went on to make the film, and Judy Garland; it is still a much-loved component of American popular culture.

WATT AND THE STEAM-ENGINE CONDENSER

The Scotsman James Watt was a celebrated eighteenth-century scientist and engineer. While a mathematical-instrument maker for Glasgow University, he was given a model Newcomen steam engine to repair. A working Newcomen engine blew steam at atmospheric pressure into a cylinder about a meter across and condensed it with a jet of cold water. Steam condenses to less than 0.1% of its volume of water. Under this great contraction, air pressure pushed the piston forcefully into the cylinder. But that jet of water strongly cooled the cylinder: it made the engine very inefficient. In May 1765 Watt was taking a Sabbath-afternoon stroll across Glasgow Green. He was musing on the Newcomen mechanism at the time and suddenly had a flash of insight. He realized that an external condenser, sucking out the steam and condensing it there, would not cool the cylinder.

Watt's mighty inspiration took seconds to have but years to make practical. Thus air always leaks into the steam of a steam engine, and a working condenser needs a special air-pump to remove it. Watt finally managed to engineer a feasible condenser. In 1775 he set up the famous steam-engine firm of Boulton and Watt.

Nowadays we think of the external condenser for the steam engine as the direct counterpart of its external boiler; their difference of temperature limits the efficiency of the engine (see chapter 15). But Watt's invention is now almost universal, and the unit of power is the watt!

TENNYSON AND "CROSSING THE BAR"

This famous 16-line poem came in a moment to Alfred, Lord Tennyson, then the poet laureate of the United Kingdom, on the Isle of Wight ferry in 1889. He was going to Farringford on the island, where he kept a house. He jotted the poem roughly down on an envelope in his pocket. That evening he told it to Nurse Durham, his housekeeper. She ran from the room, perhaps fearing that he had written his death song (as indeed he had). Later, at dinner, he showed the lines to a friend, who commented, "That is the crown of your life's work."

DYSON AND QUANTUM ELECTRODYNAMICS

This is a clear example of creativity in action. In September 1947, Freeman Dyson was a new British graduate student at Cornell University, in Ithaca, New York. At that time, many physicists were groping after a good quantum theory of the electron. Dyson, after months of hard calculation by conventional quantum methods, could reproduce some recent results. But at Cornell he met up with Richard Feynman, who had totally different ideas on how to quantize the electron. His "sum over histories" of the electron largely ignored mathematics. It exploited his physical intuitions and feelings for the electron as a physical object. He used "Feynman diagrams," a sort of pictorial shorthand, to direct his argument. He could make good predictions by scribbling on a blackboard for half an hour.

When Feynman said, "I'm driving to Albuquerque. Come along!" Dyson agreed. The trip took several days. The pair in the car bounced ideas around, and Dyson became very familiar with Feynman's way of thinking. Then Dyson went to a summer school in physics in Ann Arbor, Michigan. There, Julian Schwinger lectured on a polished and brilliant mathematical approach to the quantized electron. Dyson talked extensively with Schwinger and used his methods on several problems.

Thus Dyson's unconscious mind had absorbed two quite different approaches to the problem of quantizing the electron. He then decided to forget physics and have a total vacation. From Ann Arbor he went to San Francisco and just played around in California for about 10 days. Then he got on a Greyhound bus to return to Cornell. As they were droning across Nebraska, he had a sudden unpremeditated "moment of insight." He saw how the ideas of Feynman and Schwinger could be combined! In about an hour he had fitted the pieces together in his mind. Back at Cornell, he collected his belongings to go to Princeton, New Jersey. There he planned to write a paper expounding his insight. It took him months to get the details of that paper right, but both Feynman and Schwinger got a Nobel Prize for Physics.[12]

MY OWN EXPERIENCE OF VERY RAPID IDEA GENERATION

In 1988 I was returning from Oxford to Newcastle in my smart new motor car. I had great confidence in its powerful brakes. So I was hammering along in the fast lane of a motorway, in the pouring rain. Suddenly a car pulled out ahead of me. I jammed on the brakes, confident in their power—and kept going. The car went into a skid, not a slowing. I was going much faster than the car in front, and it was obvious that I would soon hit it.

Without thinking about it—there was no time to think—I began to pump the brakes as fast I could: on-off, on-off, on-off. I had never done that before in any vehicle, nor had I seen it done, nor had I even thought about it. At each "on" of the brakes, the car slowed slightly before it started to skid again, in that intense rain. After about a second of this fast pumping, it looked as if I might not hit the car ahead after all. I came terribly close to it. Its driver let out a great blast of his horn at the madman behind him. I got within about 50 centimeters of his car, and then began to fall back as I continued to slow. Much later I recalled the principle of the ABS braking system. By releasing the brakes and reapplying them, many times a second, it slows a car more sharply than even locked brakes could. Somehow, in that sudden emergency, I had called that knowledge up into my mind and had done the right thing. Later still I discovered that the insurance on my smart new car was not in order. Even if I had survived a fast-lane collision, legal horrors would have engulfed me. In the event, that sudden urgent demand stimulated my unconscious mind to push upstairs the crucial scrap of knowledge that saved me.

2

The Creative Environment

If we accept the theory of chapter 1, how should we increase our creativity? Creative insight may come suddenly, a wild aha! moment when you have an idea or see how a puzzle can be solved. Or it may come more gradually, a set of individual recognitions of a way forward.

I think of the single big idea as the "masculine" style of creativity, and the many contributory ideas as more "feminine." There aren't any gender implications; the creative mind should be at home with either. But a number of small creative ideas, arriving perhaps over a long time, reminds me of the long female haul of pregnancy and motherhood. Engineering designers, novelists, poets, and composers are more likely to have worked mainly in the feminine mode. Their final product may have taken years to perfect and may contain many separate creative ideas. Conversely a single powerful idea, together with the vast effort it may take to turn it into something real and practical, is in my classification more masculine. Of course, single male ideas, and contributory female ones, are both usually wrong in some way. All creatives have to live with repeated failures. Yet the few ideas that survive can lead you on to great things.

Sticking with gender notions for the moment, the business of having a new idea is rather like a woman's experience of pregnancy. A moment of great delight leads to a lot of private developments. Your distress grows; all sorts of things may go wrong; many novel efforts and activities can be needed, while many old ones have to be given up. But let us suppose that all goes well, and your new creation matures. Ultimately it hits the world in a dramatic moment of birth. Alas, the world may not be impressed.

Worse, the new creation has a life of its own. Despite your hopes and plans, it may grow up quite differently. One good example is Dennis Gabor's holography—he saw it as a way of improving electron-microscope images, rather than as a mark for credit cards.

Here is a sort of itemization of things that engender creativity. I first talk about the large-scale environment of the creative; then about the small-scale one in the creative's head. There are many such small-scale factors, and I give eight of them. Then I discuss how the outcome of creativity activates the Observer-Reasoner and then the effects of a new development on the outside world.

Environment: Large Scale

The large-scale environment includes you and the people around you. Many artists and scientists ignore it. Such "lone wolves" live and work in the traditional "room of one's own"; if scientists, they take laboratory services entirely for granted. Others are "team players." They bounce ideas off each other, have the benefits of a boss and underlings, and work as a group.

So how do groups work? Patrick Blackett (who won the Nobel Prize for Physics in 1948) said, "A good laboratory is one in which mediocre scientists can do outstanding work." He did not explain how to make a laboratory "good," and neither can I. Maybe a group of scientists can stimulate each other. Academic institutions seem better at this than industrial ones. I have to admire the Cavendish Laboratory in the department of physics at Cambridge University, the National Institute of Medical Research (NIMR) Laboratory at Mill Hill (both in the United Kingdom), and MIT. Many industrial laboratories have had moments of glory—I am thinking of General Electric in Schenectady, New York, du Pont in Wilmington, Delaware, and Imperial Chemical Industries in Winnington in the northwest of the United Kingdom—but none has lived long at that altitude. Only the legendary Bell Telephone Laboratories of AT&T was creative from its moment of inception in 1925 to its destruction (by what I regard as foolish corporate decisions) in 1984. Nothing was kept secret at Bell Laboratories, or anywhere else; but nobody was able to copy its magic. The surviving institution, called Lucent Industries, was taken over by the French company Alcatel in 2006.

The laboratory as a social institution was perhaps invented by the big German chemical companies, such as Hoescht, in the 1880s. The modern organized laboratory, with a dedicated machinist, glassblower, technicians, and analytical services, was developed in the twentieth century. Among its pioneers were Kamerlingh Onnes at Leyden in the Netherlands, Willis R. Whitney at Schenectady, and Thomas Edison at Menlo Park.

Nowadays, most organizations ponder the eternal problem "what shall we do next?" Some "research boss" decides. The decision is always tricky, for no organization has a mandate to be creative. Companies exist to sell things, colleges and universities to teach students, national laboratories to maintain standards or to produce devices for the government. So institutions tolerate creative activity as a sort of offshoot of their main enterprise.

When the fictitious organization DREADCO (a research company run by the inventor Daedalus) appeared in *New Scientist*, I was surprised to find how many readers wanted to join it. I extolled its anarchic structure, its shambolic research activities, its many unexpected successes; and numerous readers took it seriously. DREADCO, the Daedalus Research Evaluation And Development Corporation, came to acquire a lot of goodwill. The name DREADCO was invented by Edward Wheeler. That fictitious company made me think seriously about how to run a creative enterprise. I decided that much depends on the boss.

One major task of any boss is to set up challenges for the underlings. Indeed, such demands often make them creative! Self-employed creators have often felt themselves under financial pressure: a useful force that demands something but does not say what. It leaves the creative soul free to create whatever works. But let us suppose you have a boss. There are (at least) three sorts. First is the pure politico. He or she knows how to dress and whom to impress. Such a character shines in committees and meetings and rises seemingly without effort up any organization. Part of the skill of such a boss is effortless personal self-belief (for we all tend to take people at their own evaluation). The motto of this type is, perhaps, "It's not what you know; it's who you know!"

Then there is the expert boss. This type has a vast number of ideas and orders the organization to carry them out. All the underlings are frightened of such a boss. Sometimes they may even fudge the results—the ultimate scientific sin—to produce the answers that seem to be needed. If this character really knows the field and the techniques and invents schemes that often work or (since most ideas fail) can be saved if they go awry, an expert boss can keep a huge number of people busy and creative. The ultimate ambition of many of the underlings is to become a lieutenant of the Great One.

A third type of boss comes to respect those underlings. He or she spreads their achievements around and fights on their behalf if they want something. I am here thinking of Lord Ernest Rutherford, the major founder of nuclear physics. Brilliant himself, he selected brilliant students who came to dominate the field after him (a random selection might include Peter Kapitza, John Cockcroft, Mark Oliphant, and James Chadwick).

Any boss has to appreciate the ideas of the workers. He or she has to be an effective advocate. Even within the organization, this needs its own skill. Thus the chemical plastics companies have often seen their product merely as an inferior replacement. Plastics, even cheap ones in large quantities, have best been exploited by external manufacturers. One chemical company gave an employee a lot of scrap polyethylene to play with: he later left the company but took the polyethylene to start molding food containers. (This was Earl Silas Tupper, who started Tupperware.) The plastic bucket and trash can caused a lot of worry at ICI, because they could not carry hot ash like the old metal versions. I saw something of ICI's efforts to make a medical forceps out of polypropylene. The first prototype, molded at vast expense, was a clumsy copy of a steel forceps! Only at a second attempt did the company make a useful medical forceps. And that really big plastic invention, the flexible bottle that has transformed so much packaging, was never invented by a chemical company at all! The earliest one I know of was a flexible polyethylene bottle invented by a detergent company for dispensing dish detergent.

Anyway, the boss, of whatever type, has to shape the laboratory. What makes it creative? I once got talking on this topic to Margaret Steele, a massively creative person herself and the unofficial head of the laboratory mafia at ICI corporate laboratory. She reckoned that the best metaphor was horticultural. A gardener is in charge of something that nobody understands or controls—the tendency of plants to grow. The gardener can at least see that they are fed and watered, that wanted ones are encouraged, that wilting ones are attended to, that ones surrounded by weeds are given more space, and that the garden as a whole has some sort of pleasing aspect that uses the strengths of that soil and location. Furthermore, he or she understands that not everything can come up at once. The gardener will know when and how to prune or weed and when not to interfere with the plants.

The big unstated factor is love. A good gardener loves that garden. Such a one is always thinking of it, worrying about it, and wanting to make it better. Similarly, a boss who loves the subject will always be trying to improve the organization and will recruit people who love the subject too. That boss will push the organization in novel and intriguing directions that he or she loves and will contrive to build it well despite wide variations of management style.

But no boss has (or admits to having) any way of making a laboratory creative. In any case, creative people such as scientists are hard to manage. For administrators, nothing should ever be done for the first time. For scientists, such things are almost the only things worth doing. Worse, the corporate environment always wants to stop research. If the thing is not working, stop it. It is wasting company money. If it is working, stop it. Turn it into a production project, and make some money from it! Only the underling, the bench worker doing the research, actually likes it. Worse still, what do you do with a laboratory full of scientists? Somehow they have to be pushed out into the organization.

The ICI corporate laboratory was a sort of "recruiting laboratory" for the company. Scientifically minded students were taken on, soon realized that there was no company future in science, and went out to become scientifically literate salesfolk and plant managers, the knowledgeable committed company people that ICI could really use.

The people who loved science did not move out, so ICI found its laboratory slowly filling up with dedicated scientists! The administrators had no effective answer to this problem. One they tried has been used in many companies: the "scientific ladder." This rickety promotional structure is for determined company scientists. Thus IBM took on Leo Esaki (inventor of the tunnel diode and a gifted semiconductor physicist), who went up the company scientific ladder. At a high level, he discovered its disadvantages. As Jiri Janata at ICI put it, "He can do anything he likes, but nobody pays any attention to him." Ultimately he went off to a Japanese university.

And IBM tried to hold on to Gene Amdahl, possibly to keep his dangerous ideas safely in-house. But Amdahl left IBM to found his own company, making super-fast computers. In the end, his advanced ideas failed commercially.

The scientific academic faces different challenges. As a young man I spent a lot of time in the chemistry department at Imperial College, London. The staff spent their coffee breaks discussing chemistry, expounding the problems they faced and the papers they had read, or the chemical bloopers launched by their students. To lighten the discourse, there was always scientific gossip—who was in line for a Nobel Prize or to be a fellow of the Royal Society and why. Often a coffee drinker had been approached by some relevant committee and had an inside story. Then I went to the University of Strathclyde in Glasgow and found that the staff spent their coffee breaks discussing golf. (All this was decades ago and is likely now quite out of date. Everything everywhere may have changed.)

"Oddball" characters who are hard to tolerate in a corporate environment may get on better in an academic one. Thus as a Cambridge professor Paul Dirac did not want to supervise any Ph.D. student, and as a Ph.D. student Fred Hoyle did not want a supervisor. So Dirac became Hoyle's supervisor.

In some way laboratory technicians may be good judges of their nominal superiors. There is a sort of "buzz" about an active place; perhaps the technicians are always being challenged in some way. They may constantly be asked to build crazy bits of apparatus, or to prepare strange reagents, or may find a lot of intriguing samples coming for some instrumental analysis. Conversely, technicians may subtly detect a decline in the atmosphere. Jiri Janata once asked the ICI engineers about some apparatus he had designed, and they explained how busy they were. "Pity," said Jiri, "I wanted it for my garage." It was ready next day. The engineers had perhaps sensed that official demands were not always urgent.

Artists face related challenges. A major artist perhaps invents a new way of doing things. He may have several students who learn the technique and form an artistic production unit. Leonardo da Vinci said, "It is a wretched pupil who cannot surpass his master." Hence there are many paintings that baffle experts. They may be by the master or by a pupil working in the idiom the master has invented. Rubens and Rembrandt are notorious for nurturing pupils who contributed to their paintings or even carried out most of the details. It must have been hard for the master to regain control over his picture!

Environment: Small Scale

The small-scale environment is the one inside your head. J. P. Guilford divided people into "convergers" and "divergers." A convergent question has one right answer; a divergent question has many. Some people get good at convergent questions; some do better at divergent ones. The sort of challenges that I have faced, such as writing a column or devising a workable TV demonstration, have all been divergent. Even scientific research, with one answer you must get right, has a big divergent start. You have to choose the problem.

So a creative has to be at home with divergent problems. Being creative has several aspects. The first is acquiring a large mass of knowledge and experience, holding that information, and sending it downstairs to the subconscious mind. There the RIG can get at it. The second is encouraging the RIG to play around with the stuff. Then you have to stimulate the RIG to pass any resulting ideas upstairs where the Observer-Reasoner can study them. The outcome may be some sort of practical action. Both the "down" and the "up" processes are impeded by the Censor. I classify my suggestions below.

ACCUMULATING INFORMATION

The first discipline is the lifelong process of accumulating information and experience. It pays to be a "noticer." You have to develop the sort of curiosity that spots things and remembers them. Most people just discard most observations. This makes economic sense. Most of the stuff that comes in will always remain surplus to requirements. Yet I cannot imagine any way of being creative without being uneconomic and building such a mental store. You have to read and look and be curious, to be inquisitive, and to interrogate other workers, look at their experiments and study what they are up to! Copious notes help too—my "database" (see chapter 14) of stored information has been a very powerful aid to me all my life. I reckon that a lot of the contents of a retentive memory get downstairs to the RIG and aid its play. My guess is that to be usefully creative, you need at least a hundred times as much information as you will ever use.

Furthermore, as well as being curious and being a noticer, I reckon you should be an experimenter. So try things: even silly ones! Thinking with your hands, playing around, is worth developing for its own sake. You may learn something or add to the useful tricks of the trade at your command. Such playing helps you to acquire experience as well as knowledge.

A developed curiosity does not only include the things you get taught, but the subtleties few people wonder about. Why are metals strong? Why are melting points sharp? Why does plaster set hard? Why is chemical apparatus made of glass? Why do powders form heaps? Formal education, particularly the long haul that all accredited scientists go through, does such an effective job of crushing curiosity that many of its victims emerge with powerful qualifications but a fixed mental determination never to acquire new ideas or knowledge ever again. Fortunately, curiosity is quite strongly built into our nature. Even the most hidebound of us is inherently disposed to take some sort of transitory notice of the novelties that come our way. I, however, am advocating the positive seeking of new facts and chasing up chance exposures to them, quite without asking what use they are going to be. I have often roamed a library at random, pulling books off the shelves at whim and spotting notions that appeal. I have often bought old scientific textbooks, purely for their facts. Modern books are dominated by theory and only mention a fact if it has some useful explanation. But old books have lots of facts, without any explanations. Occasionally I have found things of immediate value. But the main benefit was to my RIG, which perhaps gained new playthings.

Again, I once took a microwave oven out of a Dumpster, planning to take it apart. I was just being curious. I did not know how such an oven worked and wanted to find out. I found out and also discovered that an internal fuse had blown—which is probably why the owner had dumped it. I replaced that fuse and now have a working microwave oven, which I use for chemical experiments. I. J. Good has remarked "a policeman is never off duty."[1] Neither is a scientist, an artist, or a creative.

By contrast, too many people these days claim to know nothing but rely instead on the Internet. This seems all wrong to me. Here is a seemingly unrelated example: golden syrup. This uniquely British product is a form of molasses purified to give it a light yellow color. It is sold commercially in tins and is well known to the British public. It is viscous, sticky, and one of my favorite liquids. If I happened to want its density or viscosity, the Internet could tell me at once. But personal experience is much broader and vaguer than classified computerized information. I have played with golden syrup and have the feel of the stuff in my mind. That experience has suggested to me all sorts of ways of using it (see chapter 8).

GETTING IDEAS DOWNSTAIRS: WEAKENING THE CENSOR

If you hope to be creative with your mass of data, you have to get it past the Censor. Among the Censor's jobs is that of filtering the observations of the Observer-Reasoner. It has to keep nonsense, untruths, trivialities, and worrying heresies from getting downstairs to the subconscious mind where the RIG can play with them. I do not know how severely it filters notions and observations on the way down, but I suspect it overdoes its censoring. As a result, we all lose stuff. To the unconscious mind, we must seem needlessly unobservant and incurious. A creative must try to oppose this tendency and should notice things and hang onto them, so as to give them the best chance of getting down to the RIG and being played with.

My model (see fig. 1.1) has the Censor in some sort of contact with the Observer-Reasoner. So you should be able to influence your Censor's strategy by talking to it. I like the idea of an internal conversation between the mental entities in a single skull. So I got into the habit of haranguing my own Censor. I used to remind it of the Daedalus column, which had to come out every week. That column had to be funny. After all, funny ideas are often quite close to being workable. That is why the cartoons of Rowland Emett, Rube Goldberg, or Heath Robinson are often so hilarious; the devices in them are almost feasible. Indeed, making Daedalus funny was my main problem. It was much more trouble than merely generating scientific notions with the right degree of halfbaked plausibility.

So I told my own Censor to seize any humor it spotted in the outside world. It was to grab anything that might be funny and pass it down for storage and play. A general genial interest in facts and ideas, the mental habit of valuing them for being beautiful or funny as well as for being probably right—all these mattered to me. Daedalus has even proposed (see chapter 4) that it pays for a woman to tell her unconscious mind about the sort of child she wishes to have.

Has this strategy, talking to my Censor, done me any good? I have to admit that it has never responded. Yet my attempts at getting material downstairs may have worked to some extent. I like to think that my RIG gained from having many playthings that a more censorious Censor would have blocked.

ENCOURAGING THE PLAY OF THE RIG

The third part of being personally creative, helping the RIG to play with what it has, is again a matter of personal style. I have no recommendation; we all have to find a strategy that works for us. I feel it pays to avoid routine—or if it is unavoidable, to keep changing it. A bit of time off may help too. Thus a brief holiday, or an outing to a lecture, can stir things up. Even a new environment may help. Both the Grand Canyon and the Lubyanka prison may stimulate unconscious creative thoughts!

My sense is that creativity thrives on a mixture of responsibility and irresponsibility. So it may help to be a troublemaker for your authority—devious, unpredictable, even annoying. Loafing, traveling, doodling, messing around, or trying daft experiments may also help. All these are a form of play, during which your RIG may be putting new things together. Remember, the RIG works all the time. Nobody can truly forget the job by closing the door.

CREATIVE CIRCUMSTANCES

I feel it helps to be on your own, or perhaps with a "matching impedance" (see chapter 3). Solitude may well be important for creativity. Virginia Woolf reckoned that a room of one's own was important for serious writing. The social whirl has its own conventions, which make a preoccupied creative seem odd in some way (see chapter 3). Thus the mobile phone is an ambiguous invention. By always being there, it may stifle creativity—you are never alone with a mobile phone.

Of course, the RIG is never passive. It and the Censor may push up an idea at any time, even during a party. Yet less distracting and more solitary conditions are probably better suited to idea generation. Perhaps the Censor is more permissive then. Indeed, each of us has to discover the circumstances that work best for us and learn to exploit them. Like many creatives, I have praised that dozy or reverie state, in which the Censor seems relaxed or off-duty (see chapter 1). Perhaps it lets ideas from the RIG slip past and reach the Observer-Reasoner. Maybe the unconscious mind is close to the conscious one—tears or laughter may be close to the surface too or hopes or regrets may seem unusually poignant. For me, this often happens best in bed, in the early morning. (The novelist Sir Walter Scott said the same thing; so did the mathematician Jacques Hadamard, who often woke in the morning to find that a problem he had been pondering had been solved in the night by his unconscious mind. René Descartes is said to have had analytical geometry—one of the most important mathematical insights ever made—come into his head while he was lying awake in bed in the morning.)

Another type of creative circumstance is a scientific conference or meeting. The official lectures matter less than the informal chats and arguments behind the scenes. Such chats can be more challenging or productive than routine discussions with known colleagues. Contact with a new mind often puts a new idea into the mind of a participant or rubs old ideas together. The resulting discussion may disprove some notion or put it in a new light. It may even spark a collaboration or hatch a new scheme. So on occasions these confabs may start something new and important. The conferers are usefully shaken up.

CREATIVE MOMENTS

Sometimes ideas are borne in on you gradually; but in an aha! moment one pops up with a sudden jump. I assume the Censor just pushes one up from the RIG. For me, at any rate, one accepting moment is when a project seems finished. While you are developing a scheme, you often cannot get away from the idea uppermost in your mind. But as soon as it is in practical form—a finished article, a completed apparatus, or a drawing you have put in the mail—the Censor may let new ideas up and you see quite other ways to go. You have a new notion, which may be so overwhelming that you adopt it without hesitation. So the scheme may have a rapid Mark II! On several occasions, having already designed an apparatus in my mind, I have been wrestling with the task of making a component for it when I have suddenly realized that a quite different approach would do the job much better. I have immediately diverted the project to making that different scheme.

But a new idea need not hit your mind while you are at work on its predecessor. It can come at any time. So I always carry a paper and pencil. You should too—otherwise you might have to scratch on a nearby wall, which some scientists have had to do. I even keep writing materials by my bedside, so that if I wake up at 2 a.m. with an idea, I can make a note of it before going to sleep again. Mainly, of course, this is a waste of paper; but every so often it has been important. One crucial story of this strategy is in chapter 1. I reckon it pays to make a quick note, even when the idea seems foolish afterward, as most of them do.

Aha! moments may be sudden, but they probably depend on an unconscious mental process that has grown slowly. My sense is that they build in the unconscious mind until the RIG and the Censor combine to push them up—most probably during some dramatic environmental change. Accordingly, each of us has built a pattern of life that, if disrupted, may spark a creative insight. I particularly notice how many of my stories of creativity involve journeys (see chapter 1). Journeyers are (a) irresponsible passengers, (b) deprived of their usual routine and dress, (c) deprived of their usual company, or (d) quite alone. One Nobel Prize winner has an average lifetime speed, jokily calculated by his colleagues, of 20 kilometers per hour. The illustrious German physicist and physiologist Herman von Helmholtz advocated a gentler motion—a ramble. Ideas, he said, do not come at the laboratory bench, but "during the slow ascent of wooded hills on a sunny day." I am also reminded of James Watt, whose crucial insight into the steam engine came while he was walking on Glasgow Green (see chapter 1), and of Charles Darwin, who had a vital idea for evolution theory while riding in his carriage.

Yet another possible moment of enhanced creativity can come during washing, shaving, or bathing. This may explain why people have ideas in the bathtub. I heartily agree with Benjamin Franklin who had a bath every day "not for the cleanliness, but for the thinking."

I do not know why a bath is creative, but I have frequently mused so intently in one that I have forgotten which bits of me I have soaped. Maybe the solitude of the process helps. Or maybe you discard conventional thinking with your clothes? In the 1930s, the BBC used to put its radio announcers in dinner jackets. The jackets could not be seen on radio, but it could have been that formal clothing made their statements more solemn and authoritative. And a listener to the BBC World Service Radio News has applauded it as "the truth: spoken by gentlemen."

Conversely, the steady industriousness of work seems not to stir the Censor or the RIG. Few creative moments seem to occur during routine activity. But there are always exceptions. The novelist Anthony Trollope used to set aside a specific time of day for regular writing, and he had a massive creative output! There was even a time when Daedalus had a meal in a restaurant whenever he wanted to develop an idea for a column.

EMOTIONAL ASPECTS

It is notoriously useless to argue with a committed believer. Such a person is armored against any change; I assume that the Censor and the RIG of such a one is equally rigid. Conversely, the RIG of a creative is relaxed and playful and not afraid of mental change. And since the RIG is essentially an emotional entity, emotional acceptance and lack of fear can be very important.

Thus one of the first computers in the world, the Colossus, was built at Bletchley Park in the United Kingdom during the 1940s, to help crack the German "enigma" military code. Its chief designer was Tommy Flowers of the Post Office. Unlike many of his helpers, he was not frightened by a machine with thousands of vacuum tubes. Telephone exchanges had racks of thermionic amplifiers for long-distance communication; he was used to vast numbers of vacuum tubes. This lack of fear helped him to be bold and creative and ultimately successful. On a much lower level, I have also gained from that same lack of fear. Some of my TV demonstrations showed wild ignitions and explosions: my teenage experiences and my skill as a chemist let me develop them and show them with relative skill and panache. By contrast, the TV crew and the audience were often terrified (see chapter 9).

Another way to harness the emotions behind creativity is to put yourself in some social position where you must show it. Social expectation, and the positive emotions which go with it, is then on your side. So always accept a creative challenge! You may well make a fool of yourself, but it's worth it! And offer more than you know you can deliver. Anything that stimulates the mind will do—deadlines, agreements or boasts to do something, demands for help from others, puzzles or challenges accepted, crazy projects, the search for evidence to prop up some tottering theory, anything. An ability that is exercised, challenged, and used will grow to meet the demands on it. That is why athletes go in for training. My own creativity had a special social appeal because of my weekly Daedalus column. This demanded great faith in my RIG and often raised demands for my supposedly creative services. I managed to meet many of them. But in private I was cowardly enough to hold a number of columns in reserve, in case inspiration failed or the editor objected to a column.

In particular, many scientific professionals feel the emotional tension between the changes going on in their subject, to which their own research contributes, and their status as an "expert" knowing it all. My sense is that a good RIG never rests on its laurels and is never passive. The creative mind is always turning its knowledge over, querying it, looking at it, combining bits together, seeking the next advance. The good scientists I have known have all had some sort of argument going on inside them all the time. Some people are "monomaniacs" on one topic. I reckon I am an "oligomaniac"—obsessed with a few topics, which change slowly. A strange skein of molecular and mechanical notions or arguments has been going on inside my head for decades. I am aware of the bit that occupies my Observer-Reasoner, but I would like to think that my unconscious mind is active as well. Occasionally this process reaches a conclusion, passes it upstairs, and then zooms off again.

Yet every professional gets into some main topic—mine is chemistry. I reckon that you should choose that topic at whim (and not, for example, because it pleases your parents or is likely to make money). That whim frees you to choose your subject just because you like it. That liking is probably the voice of the RIG and should make you remarkably open to all the details of that topic—both taught and untaught (see chapter 3). Your memory will cling onto them. You only need to recall that you knew them once. They exist, you know of them, and can dredge them up at need. My personal database (see chapter 14) has been valuable all my life; not just as a repository for facts, or a sort of recorder for curiosity, but as a mental stimulus.

SILLY IDEAS
Yet another way to encourage the RIG is to learn to tolerate silly questions or silly ideas. They often make us uneasy. Some of the silly

questions currently in my mind are in chapter 16. There are (at least) four sorts of silly questions. The first is a gap in one's knowledge that, once filled, fits neatly and makes you think, "Stupid me! I should have known that!"

This is the sort of silly question that nobody likes to ask, but we all need to. For example, air pressure is about 1 kilogram per square centimeter. Why do we not notice it? Air is a fluid and presses evenly: up, down, sideways, and at any angle. The human body is fluid too; it accepts and transmits that pressure evenly. So we are unaware of the pressure until we come across a different one—that of a vacuum, say, or the pressure under a depth of water.

Another type of silly question is the oversimplification of a serious claim. My favorite example is Newton's third law of motion: "Action and reaction are equal in effect and opposite in direction." So nothing can move! Or can it? It needs a bit more understanding to work out the exact sense in which nothing can move and how ordinary motion remains possible. Fortunately, you do not have to be Newton to puzzle the matter out, but this type of silly question is still worth asking.

Another type of silly question is simply odd—for example, "Can you freeze a soap bubble?" (A question my nephew asked me once. I didn't know. We tried it. You can.)

Yet another sort of silly question is a puzzle that may, or may not have a good answer; it may, or may not, be beyond current understanding. For example, I have worried how water gets to the top of a tall tree. Even the best vacuum pump cannot suck water more than 10 meters high; yet many trees are much taller than that. Maybe plant physiologists have a good answer.

If you do not accept the scientific theory of something, in all probability you just don't understand the evidence or the reasoning from it. But there is a tiny, tiny chance that you understand the evidence and the reasoning better than anyone else. In chemistry, belief in the atomic theory grew during the nineteenth century until it was accepted and taken as certain, proven, by almost everybody. But in 1904 Wilhelm Ostwald outraged the British Chemical Society with a lecture in which he denied the existence of atoms. He explained all the evidence in terms of continuous states—the physical concepts called "phases." Later came the additional evidence of radioactivity and Einstein's explanation of Brownian motion, and even Ostwald was convinced. Again, Einstein could not let go of an odd discrepancy in classical electromagnetic theory. To straighten it out, he was driven to replace the entire Newtonian view of physics. His new predictions were verified, and today we are all relativists.

I do not know any way to tell if any given silly question is profound, overly simple, shallow, or simply odd. Fortunately, Daedalus has needed a steady supply, of any kind. So I have come to value silly ideas, to make a note of them, and pass them downstairs for the RIG to play with. But they worry many scientists and technicians. People who need to "save face" must find it even harder to tolerate silly ideas. Nobody likes to "look a fool."

My brother was once in a Cretan hotel during an earthquake. He could have saved his life unambiguously by rushing out into the street. But he was in his pajamas. He quickly realized what a fool he would look wandering about the street in his pajamas: especially if the hotel did not collapse. I empathized immediately with his story—I had the same upbringing.

Not looking a fool is a strong British instinct. Having silly ideas is the same sort of risk; it takes a type of nerve even if you keep the ideas private. Mercifully the Daedalus column needed a silly idea every week. Gradually I learned to tolerate the risk.

GETTING THE RIG'S IDEAS UPSTAIRS: WEAKENING THE CENSOR

Of course, the ultimate goal of stirring up the RIG is to get new ideas upstairs: ideally in that wild aha! moment we all recognize and remember. The Censor, of course, opposes the whole process. Ted Hughes was quite pugnacious about this. He seems to have felt that to be more creative, a writer has to outwit the Censor, which he saw a sort of police enemy. I do not oppose the Censor so fiercely. Most of the time it is doing a good and necessary job. But I like the idea of weakening it, so as to get new RIG ideas upstairs. You want to get at some of the absurdities that it has been keeping down. You'll never get at personal political information (it won't let that upstairs) but you might persuade it to release more of the RIG's harmless nonsense. Even trying very hard, the Censor is not a very insightful filter. My guess is that the ideas that it lets up to be tested against reality by the Observer-Reasoner, are about 80% duds, as we saw in chapter 1. Even so, many of those duds are worth looking at.

Thus in the problem of remodeling the kitchen, the RIG might well suggest standing the fridge on the ceiling, or coalescing the fridge and dishwasher into one unit—schemes the Censor should rightly reject. But the idea of putting the dishwasher on top of a cupboard and even that of putting it inside a cupboard are worth conscious attention. They violate a needless convention that the Censor may have been applying; that every object should occupy its own bit of floor. (The best arrangement might in fact have one object standing on another.) Such needless conventions, which you may be unaware of until they are violated, curb creativity and have to be recognized and discarded. Thus I once ran away from a mechanism that annoyed my own Censor; but ultimately adopted it (see chapter 5). I got away with it, but my Censor's misgivings ultimately turned out to be justified.

Again, I have tried having a conversation with my own mental entities. I have talked to my Censor, again asking it to relax its criteria, but this time appealing to it to let notions up from my RIG! I have never had any reply, yet I have often felt that something was listening. Indeed, I slowly developed the confidence to accept a commission with no idea of how to fulfill it. I typically then said to my "entities" downstairs, "OK, unconscious mind, I have agreed to do such-and-such by this date. If you don't come up with an idea soon, we are going to make a major fool of ourselves." My conscious reaction was to look at the facts and do some experiments. But by acknowledging my unconscious mind and admitting how much I depended on it, I got it on my side, so to speak. And once I had got an idea working, I thanked my unconscious mind. Maybe it was pleased; maybe it would be on my side again, in the next emergency.

My attempted negotiations with my Censor often pleaded for Daedalian ideas. Again, I never had a detectable response; yet writing that column made me much more tolerant than most scientists are, to nonsense and to jokes. The funniness I wanted also fits Koestler's link between jokes and new ideas. So I now invite you to apply the strategy that I was trying. Feed back to your Censor that approval of funny ideas! Instruct it to pass them up without fail however absurd they may be! With any luck, your system will get more creative, you will get more joy out of your mental life, and you may think of ideas that more solemn deliberations would never have produced.

A rather general idea for weakening the Censor depends on the sound Pavlovian principle that both the Censor and the RIG, indeed the entire organism, will tend to concentrate on activities that are rewarded and to avoid those that are punished. So I cultivate the attitude of judging ideas not only on whether they work but also on what sort of human appeal they have. Thus in devising schemes for my TV shows, I rapidly learned to avoid scientific apparatus, which just frightens an audience. My demonstrations were much better as well as funnier when I used ordinary domestic objects in a crazy or surprising way.

Activating the Observer-Reasoner

The RIG may have an idea. But it is the Observer-Reasoner and its store of subconscious memories that turns it into a real-world experience and takes it further. The aha! moment when you have an idea is dramatic and memorable. But the next stages are just as important and often far more difficult. You have to evaluate your new notion—and if about 80% of ideas go wrong, you will probably have to discard it. But let us suppose that it survives. A simple experiment proves it feasible on a small scale. To make something technically practical, you may have to buckle down to a lot of hard conscious work. Even Watt's separate condenser for the steam engine, which ultimately worked and is now almost universal (see chapter 1), took years to develop. You and other workers may have to put in a vast amount of work and money. You may have to keep going back to that successful small-scale experiment, just to reassure yourself that the thing makes sense. I and many others have often had to go back to the laboratory or the calculator to renew that reassurance. Edison probably got it about right when he said that genius was 1% inspiration and 99% perspiration.

So it is not surprising that many big projects fail. Or they push the state of the art and rely on other inventions. The first computers were made in the 1940s, but the basic technology kept changing. The vacuum tube was replaced by the transistor, and the single transistor by the integrated circuit; and engineers kept on inventing new ways of storing more and more information. Thus the magnetic disk store was invented in 1967 by two IBM engineers in their spare time. When the company found out, it tried to stop them! The computer is still being improved.

Before any idea of yours develops into anything feasible, it will face the same sort of practical struggle. The thing may even get more desperate: warning you that something is going wrong. You may give up, having wasted vast amounts of work. Yet you may still feel that the basic idea was sound. This has happened to me. I can only hope that some future development, maybe in an entirely different field, will one day come down on that failure and make it easy. I shall then be an early, unsung pioneer.

In this process of trying things, I suspect that I developed a sort of practical vocabulary. It remembered things that worked, and copied them; it remembered things that failed, and avoided them. It may have evolved into a style, so that a perceptive critic could identify me as a scientist or a constructor from my creations. Some of the stories of these creations are in chapters 5 to 12. What stays in my mind is rarely the aha! moment (if there was one), but the desperate practical struggle of making the idea work.

J. E. Gordon has claimed that his designs seldom gave trouble, because he worried about them, night after night.[2] I well understand this too. The long haul of turning an idea into something practical stays with you 24 hours a day, and nocturnal worry is part of it. In my design of the chemical space "garden" (see chapter 5), I was distraught with worry for weeks before the launch. I went to the control center in a fever of anxiety, burdened by my mistakes in design—notably in the pneumatic operating mechanism of that experiment. But I knew the equipment better than anyone else; a word from me might have been vital. In the event, it mostly worked.

Even when an idea is relatively simple, and could be tested by a simple experiment, the Observer-Reasoner usually has trouble. You may have to look around at what you have got and rig something up. This part of creativity needs not only practical ability but also a sort of panache. You have the right to try something you can't prove or even something that doesn't make sense! To top it off, the experiment may annoy the authorities of your large-scale environment by using their apparatus in a funny new way. It pays to use simple equipment. This is usually cheaper, and easier to vandalize or divert to strange uses. One example is my unrideable bicycle URB3 (see chapter 5).

Furthermore, when you have turned your idea into an experiment, the Observer-Reasoner has to look at the result. This can require quite wide-ranging curiosity—especially if the results are unexpected, as with the URB3. Your attention may be drawn, not to the main reading on a dial but to some unexpected or seemingly trivial side-effect—as in my interest in the noises of steam (see chapter 12) and the hardness of crystals (see chapter 5). A developed Observer-Reasoner may spot such side effects or unexpected results and chew them over.

The Observer-Reasoner of a creative is often very acute. (One simple example is my interest in the straight sides and flat planes of crystals, chapter 16.) At its extreme, this acuity shades into fascination, which may indicate that the RIG is showing interest. You stop and stare, maybe repeatedly, and pick up even tiny cues. Thus it is said that all Americans above a certain age can remember where they were and what they were doing when they heard of President Kennedy's murder. The event clearly got into the American unconscious mind.

You can be fascinated by poetry or prose, an artwork or a scene, a material object or a person. By contrast, most writing and most people or objects are of purely transitory interest. You notice some aspect of them and gain a little information from them, and that is all. A writer, perhaps, may put thousands of words in print; but a few of them can strike you as somehow significant. Some passages stick in my mind though the ones before or after them are totally lost. A writer who does this often is, for me at any rate, a poet. The writing somehow speaks to the unconscious mind, though the conscious Observer-Reasoner may be unmoved. For example, I sense this poetry in much of the writing of the Italian chemist Primo Levi. Even in translation I feel the need to study his writing and appraise or criticize it. Further samples of language that appeal to me are in chapter 14. In some way, the Observer-Reasoner notices them and passes them down to the RIG.

And what do you do with RIG ideas that turn out wrong? Probably about 80% of them will fall into this sad category. My sense is that it pays to hang on to them, despite the trouble they can cause. Thus I once foolishly accepted a sketchy plausible argument from my own RIG and had it in my head for many years. I planned to expound it in detail some day. Only at the last moment, when I attempted a calculation to clarify the notion, did I realize that it was simply nonsense! It was too late just to admit my mistake—I had built the assertion into work for other people. I had to invent a fudged argument to give some sort of excuse for having thought that way. Yet an RIG idea is seldom totally wrong. Somewhere in it there is a core of imaginative sense for you to puzzle over, a wish that things were different, or an application in some entirely unrelated context. Or it may identify a negative principle worth bearing in mind and trying to get around. Margaret Steele of ICI never wrote a report saying that an experiment had failed—she did not want to discourage future readers. She always said that she had failed to make it work.

The Outside World

Sooner or later, you have to hand an idea over to others. The most dedicated author must give his or her masterwork to an editor or a publisher. The most dramatic work of art must be exhibited, the most powerful scientific theory must be made known, the most brilliant technology must become a manufactured product.

Then the serious trouble starts. You may get the idea in several chunks, one perhaps that starts you thinking and working and the others as you tackle the problems thrown up by the first. I feel you should praise the RIG for its contributions—but it may know anyway, from your actions. Even so, it may well take years to turn the new notion into something feasible.

Furthermore, you have to popularize your scheme so as to prepare your world. This tricky task is quite different from having the idea in the first place. I see it as a branch of advertising; it often needs a sense of drama, or a sort of jokiness. During World War II, British radar technicians used the lovely phrase of making a new circuit "sanitary." Once an idea worked, it typically existed in the electronics laboratory as a rat's nest of wires, components, and vacuum tubes. The inventor had to tidy it up, first so that the boss could see it and understand it and second to fit it for possible production.

In scientific discovery, John Ziman has pointed out the rhetorical power of prediction. It can make a new idea sanitary at once. The theorist says that something should be observed; the experimentalist goes to look, and finds it! Thus in 1705 Edmond Halley predicted, from Newton's theory of gravitation, that a specific comet seen in 1682 would return in 1758. It returned on time. Halley was dead by then, but thereafter nobody doubted Newton's theory of gravitation. Again James Clerk Maxwell's kinetic theory of gases, first published in 1859, gained enormously from its incredible prediction that the viscosity of a gas did not depend on its pressure. Maxwell himself verified this prediction—he was an excellent experimentalist as well as a supreme theoretician. During the nineteenth century, several chemists, including John Newlands and Julius von Meyer, had an idea of the Periodic Table. But Dmitri Mendeleev saw gaps in the table. He predicted an undiscovered element for each gap and boldly predicted its properties. When gallium and scandium were discovered in the 1870s, and germanium in the 1880s, and each fitted Mendeleev's predictions, his fame was secure.

The most dramatic prediction of them all, perhaps, was that of Einstein. His theory of general relativity required light to bend slightly when it passed a heavy object such as the sun. Accordingly, stars near the sun should seem to shift in position. Now you cannot see a star near the sun, except during a solar eclipse. Einstein put forward his theory in 1915, during World War I. A very favorable eclipse would be visible in equatorial latitudes in 1919. Two British expeditions were prepared (the organizers hoped that the war would be over by then). Arthur Eddington, that major pighead, was put in charge of one of them. The eclipse expeditions went out, and Eddington sent a telegram to Einstein announcing that his prediction had been verified.

In art, the only test is an audience. Much new music and new visual art was rejected by its first audience but was later accepted when its challenges were less troubling. Some was not accepted even then. I still cherish UK journalist Bernard Levin's assessment that he "would not give you fourpence a square yard for the entire works of Francis Bacon."

It is notorious that an inventor may be very poor at the development of his or her invention. An originator may not shine as the CEO of the resulting company. Reginald J. Mitchell designed the immortal Spitfire fighter aircraft of World War II, but Joe Smith controlled its production and development. Again, the computer was invented in the 1940s to handle complex mathematics. Yet it has transformed the world mainly as a communicator and as a word processor. Almost nobody now puts down words via a pen or a typewriter. And while sound recording was invented by Thomas Edison in 1877, he used a tinfoil cylinder—later a wax cylinder—and vertical "hill-and-dale" recording. Emil Berliner invented the sound disc in 1888 and used horizontal side-to-side recording. Records could be stamped out by thousands, and the recording industry took off.

Even today, despite many changes of material and format, sound is still sold on disc!

My own popularizations and TV demonstrations have shown me some of the problems of making new things understandable. Somehow you have to build a bridge that starts with the familiar and takes the audience on to the new. At his best, Daedalus did this with a scientific readership. I may have learned some of the art by "improving" the inventions of my creative friend David Andrews.

One of my most significant papers ("The Theory of the Bicycle," see chapter 5) attracted attention because it asked, and tried to answer, a question many of us had only felt—why is a bicycle stable? Again, my chemical experience of filtering wine through charcoal needed a lot of development to make a good TV item (see chapter 7). It gained from the sheer vandalism of ruining a very expensive red wine. The producer and the audience paid attention!

3

Thoughts on the Random-Ideas Generator

Creativity of any kind is rare. To be reading this book at all, you will probably have more of it than usual. But we can all hope to improve. So here are some thoughts on how a variety of factors interact with the Random-Ideas Generator, or RIG. I reckon we all have an RIG, and I at least depend on it for ideas. I am not a psychologist, so my guesses about the RIG are purely personal. But here are some of them.

Time and the RIG

How much time does the RIG need to have an idea, and how long does it wait before pushing it upstairs? Psychologists call this delay "incubation." I vaguely feel that a complex problem, or one for which the RIG has little stored data, goes with lengthy incubation.

Simple problems, such as deciding how to remodel the kitchen, seem not to need much incubation. The RIG sends ideas up very quickly; the Censor lets through those that seem feasible. The Observer-Reasoner checks them against reality. I reckon the RIG is always playing with ideas, and the current bother (such as rearranging the kitchen) is just one of them. Rapid results are usually rare. Indeed, even a few hours of delay often pays—as when you have struggled unsuccessfully with a crossword puzzle. Give up for a few hours. Writer Madeleine L'Engle recommended playing the piano as a way of "breaking the barrier." Such a complete change may help the RIG to relax. You come back with a "new mind."

A day of delay can be even more valuable. The business manager Robert Townsend recognized that a valued colleague got "negative and defensive" if pressed for an instant decision on anything.[1] Townsend advocated holding a second business meeting the next day. It allowed the slower deciders to "sleep on it." I suspect that actual sleeping is important. The unconscious mind is highly active in sleep (perhaps dreams stir it up). The physicist Frank Offner once woke up in the middle of the night with an idea about ear membranes—he was concerned with them at the time. His wife guessed what was bothering him, and said, "Now get your mind off membranes and go back to sleep." I have had the same sort of experience. Several times I have woken up in the night with the conviction that something will or won't work. Once I even went downstairs at 2 a.m. to check that something would fit! And wherever possible, I halt a project overnight. In some way the sleep experience brings a new insight to a problem, maybe allowing the unconscious mind to contribute to it. Things feel different and clearer in the morning. Some of yesterday's options now seem closed; others seem obvious.

More troublesome problems take more time. Bertrand Russell recounts his long agony of 1913.[2] He had to give the Lowell Lectures in Boston in 1914; but despite endless cogitation he could not see how to avoid counterarguments and exceptions. Finally, in despair, he arranged for a stenographer to take down a book on the topic. As she came in the door, he suddenly saw what he had to say and dictated the whole book without hesitation. So he decided that it was silly to go on worrying about a problem. "Order the work to continue underground and wait for the result to pop up," said he.

But how long should you wait? Probably Russell's RIG was sorting out the whole matter for much of 1913, unconcerned by his conscious agony. If he had hired that stenographer earlier, say after 9 months rather than a year, his RIG would have responded to the sudden challenge with what it had then—which might have been entirely adequate. Hence the value of crises. They stimulate the RIG to push stuff upstairs, even past the Censor.

My model has the RIG playing with ideas all the time. Indeed, it may have some sort of solution ready on demand. In an emergency it reacts instantly and pushes upstairs whatever it has. The idea may or may not work, but you want it *now*! This crisis reaction fits my story of Bowers in the crevasse and my own instant instinct to pump the brakes of a skidding car (see chapter 1). In my crisis of the birthday candles, my producers suddenly wanted something in a day! I had to start building at once (see chapter 10), without having any clear idea of what I was building. I developed the one idea that my RIG offered to me and managed to respond successfully.

At the other end of the scale, the longest incubation time I know of is that of the German poet Rainer Maria Rilke. In 1912 he ran out of inspiration while writing his Duino elegies. He lapsed into a frustrated depression and abandoned them until 1922. Then "utterance and release" came to him. He wrote a series of poems (The Sonnets to Orpheus) that had not been in his conscious mind at all. He also completed the Duino elegies. In 18 days he wrote about 1,200 lines of the pithiest and most carefully poised poetry ever put down. Furthermore, he did so as if taking dictation; he made very little correction. I imagine that without his conscious knowledge, his RIG and Censor had been active for years. While Rilke's Observer-Reasoner was frustrated by writer's block, his RIG was busy proposing lines that his Censor was busy rejecting.

TV deadlines can be absurdly short, and I often agreed them instantly, over the telephone ("Can you do this by Wednesday? OK—Do it!"). Indeed, this "crisis trick" can be cunningly exploited by a shrewd boss. When managing a project, he or she imposes some arbitrary deadline. The underlings get madly creative, to make the product to that deadline. At the last minute their boss relaxes the deadline. The workers can then replace their most desperate improvisation with something a bit more realistic. Their final result is more feasible but still produced in record time.

Humor and the RIG

One easy way to justify a silly idea is to say "it is all in fun." Yet any new argument may be fun or serious, or both. Take it seriously! Daedalus did this hundreds of times, of course. Thus he once imagined that a trowel, vibrating as a sound recorder, should let us recover the work songs of ancient Greek plasterers.[3] Later an American researcher published something like this for real. So maybe wheel-made pottery carries recorded sound! Again, consider the soul as the disembodied carrier of consciousness. Does it carry information? Daedalus argued that all souls must be distinct. Merely as a name or "identifier," a soul must carry 33 bits of information; so it may carry more (see chapter 16). Consciousness is still a big mystery, so this claim may not be entirely silly. Again, do dreams contain discarded memory? Daedalus has imagined that they may (see chapter 4). It probably pays to be silly; humor and creativity seem psychologically close together. I have recounted some of Daedalus's humorous collisions with the real world.[4]

Your Major Field of Study and the RIG

What makes you choose a major field to study? The minutiae of any topic can be learned or at the very least crammed into your memory, but you do well to choose the main subject of your life at whim. That whim frees your unconscious mind to choose things that seem natural or easy. Thus I went into chemistry, not merely because I was good at it, but because it appealed to me. Chemical facts seemed interesting and stuck in my mind; melting points, boiling points, chemical properties, were easy for me to learn and remember. Probably they stimulated my memory and sunk easily down to my subconscious mind, which retained them as part of the big hoard of data that I am advocating. It has stayed with me. I like to generalize that delightful usage whereby a birdwatcher, if dedicated enough, becomes a "twitcher." I am, perhaps, a chemistry twitcher.

Changing Your Field and the RIG

One rather drastic creative strategy is to change your main subject of study or at least your field of interest! Howard Gardner has recommended a change of career every 10 years. Stephen Bragg has a more exact estimate, which he has even cast into mathematical form.[5] Ignorance, he said, is very creative. When you start in a new field you have lots of ideas, most of them barmy. An old hand, with greater knowledge, will have fewer ideas but more of them sensible. If you stay in a field, you will never have a new idea. You will become a conventional expert. To use your time ideally, says Bragg, you should change fields when you have made half the contribution you will ever make. For a scientist, he reckons that in your 20s you should think about moving on after about 4 years;

in your 50s, a final slot of 8 to 10 years makes more sense. (Bragg accepts the slowing of creativity with age; I go into this topic in greater detail later in the chapter.) Thus you plan to make some contributions to six or seven topics during your career.

This seems quite plausible. Indeed, I know a woman who has changed career three times; and each time has been aware of that sudden burst of creativity at the start of a new challenge. Sir J. J. Thomson himself advised his students that when facing a new problem, they should think about it for some time by themselves. They should dream how they would tackle it, before going to the library to discover the conventional approach. The novelty and ignorance of a new mind might come up with a totally novel idea! I once met an engineer in a candy factory, whose first task was en-robing toffee centers with melted chocolate. His previous experience had been injecting oil fuel into diesel engine cylinders. Not surprisingly, he saw ways of changing the machinery of the candy factory.

Some careers (such as those of Michael Faraday, Subramanyan Chan-drasekhar, Marcel Golay, Elmer Sperry, and Isador Rabi) show lots of change. And in World War II, many British scientists dropped whatever they were doing and helped the war effort. Blackett's work for the Royal Navy, on magnetic mines and undersea magnetism, was critical to his later research. His highly sensitive magnetometers later helped him dis-prove the notion that any rotating electrical conductor (such as the Earth or the sun) should be magnetic. He turned rock magnetism into the pow-erful geophysical tool it is today.

It probably pays not to change your field too much. You don't want to leave your stored knowledge behind. A really drastic change (from physics to history, say) is probably too much. Smaller changes let those basics come with you.

Bragg's strategy seems to have worked only partially for me. It hap-pens that I have changed fields about five times, within the very broad "church" of chemistry. I started out as an organic chemist and moved through inorganic chemistry, spectroscopy, and various aspects of indus-trial chemistry, into what is probably now chemical physics. I had no subtle Braggian motive; it just occurred (fig. 3.1).

I have never felt that burst of "new field" creativity. A new field cer-tainly gives the RIG a lot of new problems to play with. But in my sad experience, you spend the first year or so doing daft things, failing to take

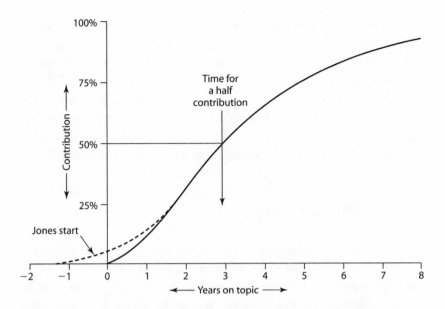

FIGURE 3.1   Stephen Bragg's Theory of Contribution to a New Field
Bragg (*solid line*) reckons you should change fields when you have made half
the contribution that you will ever make. My own dotted start to his graph fears
that for at least the first year in a new field, you are simply incompetent.

obvious precautions, not knowing elementary techniques, or following
silly notions that those already in the field know are pointless. Thus in
the third year of my Ph.D., I repeated all the work of significance that
I had done in my first year. I lost two weeks to the effort, and half of it
was wrong.

But it still may help to have two entirely different problems running
together. My serious research was probably a useful counterbalance to my
journalism, my eccentric personal projects, and my mad popularizations.
That serious research usefully counteracted the crazy world of Daedalus,
while benefitting both. Bertrand Russell praised the philosopher Baruch
Spinoza, who was also a lens grinder. The lenses made a useful task when
Spinoza was stuck with his philosophy. Furthermore, it often happens
that an idea comes in several stages. The great Isaac Newton presumably
wondered about the falling apple and the motion of the moon as separate
problems in physics. Then he connected them into one!

Lesser minds may also have gained by making connections. Several times I have had a "fractional idea," which made no sense on its own, but somehow had potential. Later, maybe years later, another bit would surface, and the combination was then usable. Thus Daedalus's "bubsub" submarine combined a tent to let a hamster breathe under water, which I read about casually in 1964, with a toroidal bubble I glimpsed in my bathtub in 1981.[6] My sense is that you should have a lot of room for mental junk, just as you have (or I have) room for physical junk. Every so often, a bit comes in handy. It pays to be a hoarder!

Social Skills and the RIG

Eric Laithwaite was a professor of heavy electrical engineering at Imperial College from 1964 to 1986. He was one of the main brains behind the linear electric motor as a device for high-speed heavy transport, needing no wheels or rotating elements. Indeed, he invented the "magnetic river" linear motor for heavy vehicles; it combines drive and lift. He mused much on the philosophy of electric motors and that of invention.

He had great social skills and gave many lectures at the Royal Institution. I have from him the phrase "talking into a matching impedance"—a notion he got from his Ph.D. supervisor and took over himself. A matching impedance is some person, who you do not merely like talking to (indeed, you may get angry) but who in some way encourages you to bring out your ideas. Maybe the matching impedance helps overcome the Censor. It doesn't matter if they know anything about the topic; indeed it can help if they don't. Eric once said, "I have to put things more clearly for the matching impedance than I would for myself; and in that clarity I say something different." He claimed that of his many inventions, he had all but one while talking in this way. I also have known "matching impedances," listeners who could convert me from a reticent loner to a flagrant exhibitionist.

Collective Creativity and the RIG

Alec Osborn, the originator of the collective "brainstorming" session, has said that the central principle applies just as well to individual ideas. That principle is "deferred judgment" of an idea. In other words, hang on to an idea that just seems silly nonsense. It may later make sense or may spark an improved idea. Thus in a brainstorming session, a group of sailors imagined a mine drifting near the ship. Somebody said, "Everyone go the side and blow!" This silly idea led to the notion of turning the ship's water hoses onto the mine, pushing it away without setting it off. This book is not about collective invention but rather individual creativity. In all my life I have encountered just one invention that seemed a collective product. It was the O-ring that damped possible vibration of the reaction cell in the chemical space "garden" (see chapter 5). I cannot say whether I or the engineers Roland or Bruce thought of it, though we were all agonizing over the problem for a couple of weeks.

Yet I do not want to discourage collective creativity. A good listener has a rare and valuable skill. Creative duos like Gilbert and Sullivan or Harburg and Arlen (see chapter 1) may have worked because each was a good listener to the other. If you know a good listener, cultivate that acquaintance! C. P. Snow has claimed that British prime minister David Lloyd George was a splendid listener; so perhaps was the chemist and writer Primo Levi.

Art and the RIG

Many writers have claimed to have an inner "daemon," whose composition must be respected. J. B. Priestley and Rudyard Kipling, for example, have written about their daemons.[7] I suspect that most great artists have one. I am inclined to identify it with the RIG of the good scientist. It seems to be some creative entity in the unconscious mind; Kipling in particular advocated walking delicately while it was at work in him. We probably all have one—but many of us let the conscious mind shout it down. I reckon that Kipling's policy was better.

A Stream of Ideas from the RIG

If you ask a number of engineers to list possible improvements to a process, their lists are likely to start much the same. Only far down the lists will different engineers start making different suggestions. This suggests to me that the engineers all maintain a list of simple ideas to trot out to any manager who demands one. Only if you keep plugging away do you start to reach their new and original notions. I reckon the RIG has the same sort of approach. If some modest scheme can get past the Censor and satisfy the Observer-Reasoner, splendid. Hence, perhaps, the first idea that comes into your mind may not be the best. It is probably one of the dud 80%. It pays to discard it, and demand another and another and another. After a while the RIG begins to take the challenge seriously, and it may push up something truly bold. This may be relevant to the feminine style of creativity (see chapter 2). Thus I once struggled with a rather poor scheme for a Daedalus column. I thumbed through textbooks for hours but still could not make the idea work. Then quite a different idea suddenly popped into my head. I saw at once that it was much better than my previous one and wrote an entirely different column. I now reckon that my RIG had realized that its original scheme was not working and pushed up another.

Drugs and the RIG

One obvious trick to stimulate your creativity would be to use drugs. It would be wonderful if a specific drug activated the creative mechanism without bad medical side effects or the possibility of addiction. Sadly, I know of no such drug, certainly none that seems to stimulate a lethargic or undeveloped RIG. If drugs aid creativity at all, they do so by speeding up the time-scale of the unconscious, perhaps simulating the crisis mechanism that encourages the RIG to push material upstairs. Or maybe a drug can sabotage the Censor. Some people are normally so uptight that they need chemical help to be creative at all. Otherwise they dare not let anything upstairs. In the same sort of way, some people have never made love sober: they can only "let go" drunk. I am sorry for them. But does that simple drug, alcohol, help? Many writers have been heavily alcoholic. Would they have been better writers sober? Is alcohol more easily tolerated by writers than (say) by physicists? The poet A. E. Housman used to have a glass of beer at a pub during lunch. He said that it helped to sedate the intellect; his idea was to go for a stroll in the afternoon and hope to compose some poetry. The fictional British journalist Lunchtime O'Booze would have approved.

My guess is that unshielded exposure to the ravings of the RIG may be a useful blessing in art, where almost anything goes. In science and technology, where almost nothing goes, it is liable to be counterproductive. I once got so drunk to write an article that I neither corrected the spelling nor altered a line—but I did not repeat the experience. I later decided that I did not want to write for any magazine that would accept stuff like that. Raymond Chandler was once made drunk by a producer who wanted a film screenplay finished on time. Nobert Wiener, a mathematical genius, may also have been an alcoholic.

Other drugs have a limited record. Some jazz improvisers have attributed many of their spontaneous inspirations to cannabis. Paul Erdös, a mathematician so strange in all other ways that he makes a bad example, took Benzedrine regularly. He said that it helped his concentration, and he could tolerate the side effects. Hunter Thompson (the "gonzo journalist" who wrote *Fear and Loathing in Las Vegas* and killed himself, aged 67, in 2005) was some sort of extreme case of drug use. Thomas de Quincey, a nineteenth-century writer, was an opium addict, and so was the artist Aubrey Beardsley. All these people were pretty talented even stone cold sober. It is not obvious to me that their inspiration relied on their drug, although they may have valued its accelerated time-scale in meeting deadlines.

One artist has told me that he had done some splendid work on a LSD trip but used several felt-tip pens in different colors to get the chemically inspired notions down fast, because he knew another would come soon. One counterexample is Francis Thompson, the Victorian poet, who gave up an opium addiction before writing his most famous works. A drug that specifically stimulates the creative mechanism does not seem impossible, but I do not know of one (yet see chapter 4).

Eccentricity and the RIG

Does the creative or original mind have such internal troubles that its owner outrages social norms or seeks medical help? Not obviously. John Dryden has famously written "Great wits are sure to madness near allied, and thin partitions do their bounds divide." He was probably recalling the Roman writer Seneca: "Nullum magnum ingenium sine mixtura dementiae fuit"—No great ingenuity is without a mixture of madness. Dryden's claim is widely quoted, but the facts do not seem to support him. For this book, I have looked at the lives of many creatives and have noticed no useful correlation. Obvious eccentricity seems not to help. Many eccentrics are in fact "licensed lunatics," tolerated by society but still carefully self-aware in conduct and opinion. Numerous studies of creatives have tried to identify some "marker" of future creative achievement. They often posed questions like "how many uses can you think of for a brick?" I like to imagine a young Darwin or Einstein struggling with that one. Yet one finding does encourage me—the educational achievement of a future creative is often mediocre. Many eminent people did badly at school—maybe refusing to work hard at subjects that did not inspire them or annoying their teachers by proposing unusual answers to questions. Many of us have mediocre educational achievements, so maybe many of us are potentially creative!

David Andrews once remarked of an invention of mine that "it takes a really warped mind to think of something like that." On that criterion, his own mind was even more warped. Yet Edison scorned "long-haired types," saying that they never made worthwhile inventions. Maybe he was dismissing his licensed lunatics. Havelock Ellis felt that less than 5% of British geniuses seemed insane, as the term was used in 1904.[8] This makes me think of the nineteenth-century mathematician Georg Cantor. He invented the theory of infinite sets, which finally sorted out the thorny problem of infinity. Cantor spent much time in an insane asylum. My own sense is that it takes talent to understand a new form of reasoning, but genius to invent one. The neighbors or the medics are a secondary problem. The true genius is driven by internal need. Einstein comes to mind, with his drive to comprehend the physical laws, "like a man gripping an object in his fist" as C. P. Snow has said.

A creative type is often odd in some way. He can be denounced as arrogant, selfish, or ruthless, when in fact he is simply absorbed in a problem. Despite the people around, he is mentally alone. This preoccupation may make the creative thinker seem antisocial. Persistent musing on internal ideas fits the social scene badly. Yet the historical eminents mainly had enough ego strength or ego resilience to combine a novel internal life with adequate social skills. I have read that many of them were "stable introverts."

As for mollifying the neighbors, most people can manage this without detracting much from the main thrust of life. It helps if you are extremely rich, as was Henry Cavendish (1731–1810). The family mansion was in

Clapham, near London, and he turned it into a workshop and laboratory. He had many servants but was so antisocial that he never wanted to see one. They had to keep out of his way! He even found meetings of the Royal Society an almost overwhelming trial. And yet he had enormous skill and patience as a scientist. He spent years establishing that water was composed of two parts of hydrogen and one part of oxygen, when these two gases were both very hard to prepare, identify, and purify. He also made many analyses of air. By 1781 he had found a small proportion of the air, about 0.7% by volume, that was too inert to be any recognized gas. In 1895 Lord Rayleigh, who discovered argon in the air, announced that air has about 0.9% of argon by volume. (Atmospheric argon might help us save heat, see chapter 15.) We have only one portrait of Cavendish. The artist had to sketch it quickly before its subject found out. He was highly antisocial but a true genius. The Cavendish Laboratory in Cambridge is named for him.

Most people are better socialized than Cavendish, but most are less creative. Some are so socially skilful that they rise to the top of any organization that employs them, no matter what they think or how little they know (many of us suffer under bosses like this). Others are so awkward, or are such pains in the neck, that almost no amount of genius can save them. The physicist Stefan Marinov may be an example; he was so ill-equipped for the world that he killed himself in 1997. (For his literary style, see chapter 14.) He antagonized all other physicists, sneered at them, and may have been crazy much of his life. He claims to have dug the grave of relativity in the prisons and psychiatric clinics of Sofia (the capital of Bulgaria). And yet his scheme to determine the speed of light, not just both ways along a reflective A-B-A path, but along A-B and B-A separately, ought to be properly done. Marinov's own attempt to do it, in his girlfriend's apartment, annoyed her so much that she threw him out, along with his apparatus.[9] Even with a girlfriend, his interpersonal diplomacy was terrible.

Gender and the RIG

The history books salute mainly male creatives. By contrast, women are largely disregarded. Are we ignoring, or making things hard for, one creative half of the human race? Men and women are different, of course.

Modern political correctness requires us to pretend that everyone is the same as everyone else—nonsense that will be finally laid to rest when each of us has our DNA individually deciphered. There is, of course, no reason why the sexes should be identical. They have many differences, each of which must have an evolutionary purpose, whether we understand it or not. For example, women typically live longer than men, which complicates many pension schemes.

In my own research on this topic, I have studied the IQs of men and women. Even if the averages are the same, the spread among men seems to be greater than among women. Thus a graph drawn from IQ data, shows that the stupidest people and the cleverest people are likely to be males (fig. 3.2). Godfrey Hardy has pointed out how useless intelligence is by itself. I do not know what addition he might have suggested for high creativity. A retentive memory? A determination to do something important? A passion for some subject? A drive to convince others? I have read that above an IQ of 120 (not all that bright; the average is 100), there is very little correlation between intelligence and creativity.

In chapter 2, I contrasted the "masculine" style of creativity, with one big idea, with the "feminine" style of many unrelated ones. If creativity is distributed among the population like IQ is, it might be dominated by rather odd males (as indeed seems to be the case). Yet I like the idea of creative women, too.

Years ago there was a lot of talk about "feminine intuition," which was roundly criticized by feminists. It has also been brilliantly denounced by Alan Sokal and Jean Bricmont.[10] My own vague theory at the time was that the traditional role of women in society imposed a sort of passivity on them. So they assessed the people around them by shrewd private observations, which were seen by many as "feminine intuition." (I discuss intuition further in chapter 4.)

I would expect the male and female creative styles to differ but have no ingenious way to tell them apart. In chapter 2, I claim that enterprises that take a lot of time, and that depend on several independent ideas, are naturally "female." Thus it takes ages to write and polish a novel, and many of the great novel writers have been women. Similarly, a poem can take years to perfect, and many great poets have been female. A female creator may nurse an inspiration for years, overcoming the problems as

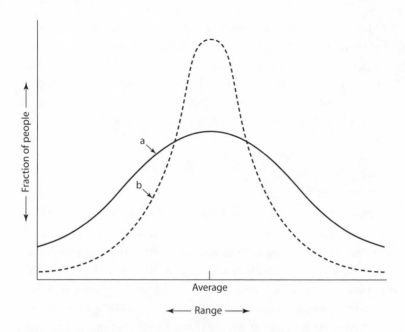

FIGURE 3.2   Graphs from Data on Male and Female IQ
Line *a* is for males. It implies more male idiots and more male bright sparks.
Line *b* is for females. It implies a tighter spread about the norm. Note that the
average is the same for both.

they arise. (Come to think of it, I did this with the steam balloon, and it
took me a lifetime; see chapter 6.)

But once an idea or an intuition is "up," it is in the province of the
Observer-Reasoner. It may be a private suspicion about an individual; but
often it must be conveyed to other people. And the proportion of female
ideas that make public sense is probably much the same as for male ideas,
perhaps 20% at best. Male and female alike, we all have to "kill our babies," or most of them anyway. Furthermore, it is your duty to test a new
idea to its limits. You must expose it to criticism, get it studied by bitter
enemies, and—the crucial test—try it out in practice. It may take years.
But Mother Nature is cleverer than any critic, and she has been around
longer, too. She is the ultimate umpire.

Art is even harder to judge. It depends very much on intuiting the
reaction of an audience. Further, the product may be in "advance of its
time." It may appeal more to an audience that appears after the artist's

own lifetime. Indeed, one searching test of a work of art is its ability to appeal to audiences far removed from the artist's experience. Don't ask me to guess whether male or female products will be more enduring! My only advice applies to any creator of anything. Go away and come back later, when your immediate enthusiasms and assumptions have decayed. Almost certainly, you will see ways of changing or improving your creation.

Sexuality and the RIG

Around puberty, three things (at least) happen to the human mind. Your interests in sex, and in physical dominance (or the human pecking order) rise markedly, and so does your creativity. All these may be due to the sexual hormones that enter the brain at that time. Indeed, Daedalus has put forward this notion quite boldly; I discuss his claims in chapter 4. Here I merely suspect a connection between sexuality and creativity. The main evidence is that sexual interest, like creativity, rises rapidly at puberty, but then declines slowly over life. Successful sexuality, of course, requires you to negotiate with others. Many people find this hard and respond with depression (which I connect with creativity below). This notion does not distinguish between homosexuals, heterosexuals, men or women. I have seen the claim that much of human activity—language, art, humor, hunting, poetry, militarism, crime, dancing, sport, politics, ideology, music, song, mathematics, exploring, philosophy, science, technology—evolved as male displays to impress women. They are all disguised forms of sexual appeal.

The impulse to mate begins at puberty and may help the initiation of creativity, but it may not be conscious even then. An interesting example is that of Charles Babbage and Ada Lovelace in the nineteenth century. These mighty pioneers imagined the computer long before electronic technology. Babbage's "Analytical Engine" used mechanical gear wheels, and he never managed to make it. But Ada conceived the first programming language for it. One hundred and fifty years later, an electronic programming language was named ADA in her honor. Nowadays it is easy to imagine that Babbage and Lovelace were secret lovers; but they were probably quite unaware of this aspect of their association. Their joint creativity was just one of the mighty achievements of the British Victorian era. My suspicion that unconscious sexuality was part of it, and of the huge creativity of that whole time, is simply a guess.

Anyway, creativity may be related with sexuality, if only in that both start suddenly and then decline slowly with age.

Age and the RIG

Age subtracts from all our natural abilities. The poet W. B. Yeats wrote the despairing couplet, "Who could have foretold / That the heart grows old?" He was so upset by advancing age that at 68 he underwent the Steinach operation, intended to boost genital capacity. I do not know whether Yeats hoped for renewed poetic creativity as well. Modern psychologists reckon that the operation has no mental effect, but it may have given him a subjective lift.

One piece of statistical evidence about age is the death rate. This differs for men and women and seems to support male creativity. Slightly more males are born than females, which compensates for the slightly greater tendency of males to die younger. So the sexes reach maturity in about equal numbers—itself evidence that monogamous coupling is the human norm (fig. 3.3).

For both sexes the "actuarial prime of life" in which you are least likely to die in the next minute, is about 9. But just about this age an upsurge starts in the curves, and the sexes diverge markedly.[11] There is a bump which I call the "curve of youthful folly." It is bigger for males than for females—most "youthful folly" is male. Is the sort of escapade that results in death related to the exercise of creativity? I am guessing so. To me, this shows that women are often more sensible and consequently less creative than men are.

Above the age of 40 or so, the death rate fades into a logarithmically linear section, when perhaps both sexes have done with youthful folly. Thereafter the chance of death rises evenly and monotonically. In age, women seem to wear somewhat better than men. My curve guesses that without youthful folly, the death rate of the young would be even lower.

Most creative art and science has been produced by young people. Among the Nobel Prize winners, the greatest age for a major physical achievement may be that of Dennis Gabor, who had the idea for holography at age 47. Max Planck founded the revolutionary quantum theory

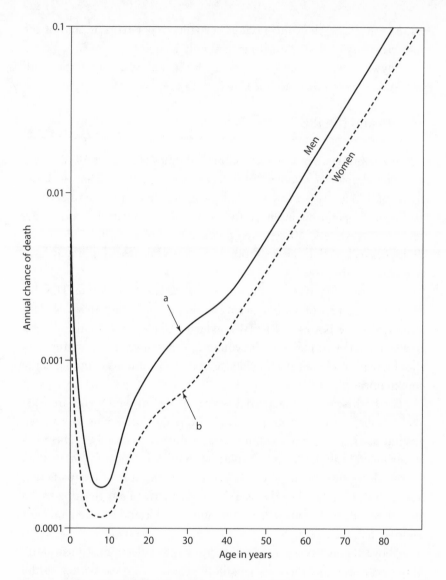

FIGURE 3.3   Death Rates for Males and Females
The bump, which I call "the curve of youthful folly," is larger for men (*a*) than for women (*b*). I interpret this graph as implying that young men are somewhat more creative than young women and are therefore more liable to indulge in crazy exploits that end in death. I smoothed the curves above from published statistics on the deaths of UK citizens during 2000. They start at age 0 with about a 1% chance of death (this is mainly the mortality of being born) and decrease rapidly as the child matures. But above the age of 10 or so, the death rates start to climb. The sexes diverge; in general, women last longer than men.

when he was 42. His personal life was tragic, but he changed for ever the way we all think. Godfrey Hardy noted that "Galois died at 21, Abel at 27, Ramanujan at 33, Riemann at 40. I do not know of a major mathematical advance initiated by a man past 50."

Elderly scientists sometimes "go peculiar" in the judgment of their peers—I am thinking of Linus Pauling and his advocacy of large doses of vitamin C and of Fred Hoyle with his interstellar bacteria—but this is all part of the aging process. You may not get a generally accepted answer, but at least you are being creative, thinking and questioning, looking at evidence, and keeping your mind active. And you may be right!

By contrast, many young people are naturally highly creative. Their creativity is often a powerful counter to their limited mental store of knowledge. (Some of my teenage ideas came true decades later; one features in the Toyota Prius fuel-saving car. And in mathematics, where things exist merely by being defined, ignorance is a minor bar. There have been many brilliant young mathematicians.) One artist, while a teenage student, filled a lot of sketchbooks with creative ideas for pictures. He kept them. When he drew those early sketches, he lacked the artistic competence to make pictures from them. Over his working life his creativity declined, but he gained skill as a working, practical artist. For him, those teenage sketchbooks became an increasingly valuable resource. He could now use their ideas!

Most of us build over our lifetimes an ever-larger store of knowledge, experience, and detailed expertise. It probably counterbalances our slow loss of inherent creativity. Eminent creatives have felt little internal losses between (say) 50 and 80. Much depends on health. Yet all of us slow down with age. I guess that the "incubation time" for inspiration lengthens, and it can help to wait a bit. Yet even without trying, we gain knowledge: "know-how," the "tricks of the trade," a "sense of the business." I have read that the financial collapse of 1987 was made worse by young whizz-kids who had never known a falling market. They had devised clever computer programs to exploit the market but had omitted to fit them with stop-loss provisions against a fall. Older, wiser heads with more experience were not fazed by a falling market but did not understand computers.

Performance art may benefit even more from experience. Youth may perform with mechanical perfection, but age can give magical expressiveness. In any case, the creative decline with age is completely statistical. Many people defy the statistics and are far more creative than the norm. Grandma Moses began to paint at 75, and Benoit Mandelbrot has said that "mavericks" can present new ideas at a great age.[12] So no matter what age you are, or what sex or social standing, keep stimulating your creativity!

Depression and the RIG

My model suggests that artists, scientists, jokesmiths, and creative people generally, would tend to overlap, or at any rate would be psychically close together. This is not obviously true. The one thing that they have in common is also common to all humanity: they all spend a fair amount of time being down. Yet depression generally remains a personal story; first because it has seldom been advertised by the sufferer, and second because it is often not considered a disease by the doctor. In any case there has never been a swift and reliable treatment.

Human beings have evolved from animals very rapidly—in a few million years only, which is extremely fast on the biological scale. We still have lots of snags in the hardware and bugs in the software, such as our vulnerability to depression. Why has evolution not just eliminated it? Unlike many diseases, it seems not to be caused by an invading organism. One variant of it is bipolar disorder, or manic-depression. Spike Milligan, who changed the whole nature of British humor during the 1950s, was a manic-depressive. He once asked his secretary to shoot him. She declined. It may have been outside her professional duties. Other comedians often ghostwrote for Spike when he was unable to write himself. He was massively grateful for medical lithium when it became available. My guess is that the new and mighty human brain has many inherent disadvantages, and evolution is still trying to optimize it. Depression and manic-depression are among its attempts to combine the power of creativity with that of rationality. This is not even a theory—it is a sheer guess. But it may make some sort of sense. I vaguely think that the unconscious mind creates new ideas during the depression, and some of these get upstairs during the mania. Kay Redfield Jamison has discussed many manic-depressives among nineteenth-century writers and artists.[13]

My remarks on depression do not distinguish between bipolar disorder and the more common unipolar style of depression. Either can induce the utter wretchedness that Lewis Wolpert has called "malignant sadness." Such an extreme of despair needs psychiatric support; and I applaud the modern antidepressant drugs that give the psychiatrist new and welcome powers. But we are all familiar with a less extreme state of being down. It may even be more pronounced in creative types. I have had attacks of depression myself—in chapter 13 I note how dramatically my weight declined during such an attack, and how it recovered afterward. I can also recall having to create a funny joke at a time when I did not feel at all funny. It was one of my best! I have been encouraged by a saying due to Alistair Cooke: "The professional can do his best work when he doesn't feel like it." Kipling has a poem in which the royal jester Rahere gets depressed—perhaps an occupational hazard for any jester.

The poet Stevie Smith once wrote that she only listens to her Muse when she is unhappy.[14] I take the Muse here as being her RIG. If so, her verse seems to support my theories about creativity. Smith feels powerless to create any writing by conscious intention, but depends on her Muse to do it. Again, Mike Adler's intellectual diarrhea (chapter 1) may have been a sequel to postviral depression. It further suggests that the RIG is churning away much of the time but can only get its ideas upstairs when the Observer-Reasoner is in a fit state to accept them, that is, unhappy.

I guess that when you are down the unconscious mind is very active, perhaps stealing mental energy from the conscious Observer-Reasoner for its own use or diverting time and resources in the brain. Thus after separation or bereavement, an individual is often in a state of mental fatigue for a year or so. During that time, the unconscious is redrawing the secret mental wiring diagram of known people and what you really feel about them. All that "personal political information" now has to be rejiggered to fit the new realities. It is a big job, and takes a long time. The brain is a slow organ at changing stored information (though it specializes in the fast recall of existing stored information). Unlearning is always very slow and difficult.

So I suspect that depression may be a sign of unconscious activity. Stick with it; perhaps something important is happening! Thus I was strongly depressed in 1966 and 1968, during which years Daedalus had two of his most brilliant ideas—the hollow carbon molecule, which won the Nobel Prize for those who finally made it and the chemical laser weapon that later became part of President Reagan's "Star Wars" project that was made real by the U.S. Air Force for the Cold War politics of President Reagan.[15] I can recall dreaming up that laser, thinking that I was playing a joke on the "energy level" notion. Unknown to me it was seriously possible and was already being worked on. But I published first!

I am also thinking of the mighty creative physicist Wolfgang Pauli. He was a caustic individual, known to Paul Ehrenfest (another great physicist) as "God's whip." I once met a scientist who had been an undergraduate under Pauli and who had ventured a possible solution to a physical problem. "Wrong. Quite wrong!" snapped Pauli. My interlocutor accepted this as a fairly mild Pauline rebuke. It certainly compares well with Pauli's dismissal of a physics paper he disliked. "It's not even wrong!" he said. Victor Weisskopf took his Ph.D. under Pauli and later became a powerful physicist in his own right. He remarked, "It was marvelous working with Pauli. You could ask him anything. There was no worry that he would think a particular question was stupid, because he thought *all* questions were stupid."[16] (My own subdivision of stupid questions is in chapter 2.) Pauli is perhaps the archetypal scientific pighead.

Pigheadedness is the certainty that you are right: the blank refusal to be influenced by your peers. It is some extreme form of dominance. It is well defined by one of Bertrand Russell's irregular verbs: "I am determined. You are obstinate. He is a pigheaded fool." In science, Albert Einstein was a major pighead; he rejected the whole statistical basis of quantum mechanics. (C. P. Snow called him, politely, "unbudgeable.") A random selection of other candidates might include Wolfgang Pauli, Arthur Eddington, Fred Hoyle, J. B. S. Haldane, Cyril Burt, John Maynard Keynes, Barnes Wallis. If you are wrong as well as pigheaded (as some of these were, some of the time), you don't cut such a fine picture in the history books. At least you may have saved others from a fruitless exploration. And your time may come later!

Pauli won the Nobel Prize for physics in 1945, essentially for Pauli's principle, which makes sense of much of molecular structure. He is now best known, perhaps, for his prediction in 1930 of the tiny neutrino particle, 30 years before it could be observed experimentally. One school of thought now holds that much of the universe is made up of neutrinos.

Pauli became depressed in 1932; it bolsters my theories that such a powerful and creative individual could become depressed. He went to the great psychiatrist Carl Jung, who offered to study his dreams. My fantasy is that the Nazis had bugged Jung's consulting room, so that somewhere there is a transcript of the conversation between the two Greats. Jung would certainly have found Pauli's dreams interesting. I suspect, however, that they contained very little physics. Even Pauli's dreams were probably personal politics, heavily disguised.

4

Intuition and Odd Notions

I mentioned intuition in chapter 3. I reckon that either sex can develop it, and it matters in creativity. In particular, I admire the "physical intuition" of the good experimenter or engineer. As I see it, intuition is not just inspired guesswork but derives from observation. It may be a subtle sort of "pattern recognition," the trick by which you identify a face or an expression on one. Applied to humans, it probably comes from seeing the tiny cues that most of us give out. We learn to "read" those cues while we are growing up. Indeed, knowing them is almost the only defense a child has against adults. The psychiatrist Eric Berne has extolled that childish skill. He has called a child a "little professor."

Intuitive skill is not innate or limited by gender—it can be learned and developed by anyone. Everybody, male or female, can have intuitions. In my view, they are very special and original observations, seized by the Observer-Reasoner and held in the subconscious mind. In scientists or technicians, the result is "physical intuition." Physical intuition can be very important to a creative, especially one who works with the material world. It can dominate one's practical style. It is, perhaps, the ability to invent a physical experiment or guess how one will go or to see that something won't work or is worth trying. I reckon it compresses a vast amount of remembered practical knowledge and observation, acquired by looking with real curiosity at every aspect of many experiences. Small observations that most of us discard, such as the feel of fitting a nut onto a bolt, or seeing that two powders have a different shade, may get into the subconscious memory of an intuitive.

A good experimenter accepts this intuitive feeling and builds on it; I think of it as a sort of practical flair. It seems unconnected with a deep theoretical grasp of the subject; indeed, theoreticians often do not have it. Margaret Steele at ICI had it at genius level. She kept on with many experiments that I thought were silly, varying them until she had something obviously important, even to me. Physical intuition may reveal a person's love for the small signals of physical reality. I can perhaps spot this sort of intuition and admire it greatly. I may even have it to some extent myself. Thus if I am asked something, or find myself wondering about something, and it is quite beyond me, I make a guess. It exploits my physical intuition and may be quite good. How long does it take under vacuum, for the gas in the bubbles of a foamed polystyrene ceiling tile to diffuse out of it? Before making a test, I guessed an hour; the half-life is in fact about 40 minutes. Can you make a steam balloon from a trash bag? Yes, you can (for my lifetime of struggle with this notion, see chapter 6). Can you make a good optical mirror by sucking at aluminized plastic film? No, but the result can still be useful (see chapter 10). Each of these is a physical intuition.

In the early stages of a project, physical intuition may push you this way but not that way, or may advocate this option but not that one. My TV colleagues have commented that when they first discuss a topic with me, I may propose many ideas. As the project develops, most are rapidly abandoned, part of the rejected 80%. But when physical intuition works (as in the brilliant solution-paste for growing a chemical "garden" in space, chapter 5), it can be far quicker and more original than traditional development.

The Observer-Reasoner of an intuitive must notice a lot of observations and pass them down to the subconscious mind. They may be very small (as in the "pit sense" of a miner, the navigational skills of a Polynesian seafarer, the subtle judgment of a Mississippi pilot). Probably the Observer-Reasoner of a nonintuitive person does not notice or note such little observations; indeed it may discard them totally. Even an intuitive person may be quite unaware of how many of those small observations are being retained in memory. Consider, for example, the "pit sense" of a miner. This is interesting because mining was traditionally a male occupation, so that pit sense was usually a male intuition. It warned the miner of sudden danger. A tiny cue—the creak of a pit-prop, a minute shift in a rock face, a slight change in the smell of the air—warned the miner to take immediate defensive action. Its promptings could strike at any time, no matter what the conscious mind was up to.

A good example of intuition, its independence of normal reasoning and its exploitation of very tiny observations, is the story Konrad Lorenz tells about the parrot Geier.[1] Geier said "Auf Wiedersehen" whenever a human guest departed. No faked departure could elicit this response; but a real one, however unobtrusive, always did so. Lorenz and other leading behaviorists failed to discover the cue that Geier was reacting to. It may have been a very tiny human tic or some distributed style of behavior. Whatever it was, that tiny "departure" signal triggered Geier's intuition and provoked its vocal response. Maybe this cue was spotted by an alert avian Observer-Reasoner or maybe it got down to Geier's subconscious mind.

Human intuitions need not depend on tiny observations; but they can still lead the Observer-Reasoner in a totally new direction (see the discussion of the unrideable bicycle URB3 in chapter 5). I advocate encouraging and giving in to any sudden strange "whim" that hits you during a rational exploration. It may be a snatch of physical intuition!

Eric Berne has discussed intuition: for example, his intuition of the age and region of origin of a person.[2] Berne has sometimes felt that his own intuition was good. I have never had that feeling but still regret having ignored internal intuitive warnings. Neither of us knows where intuition comes from, though animals (and even some human beings) seem to rely on it entirely.

Applied to the material world, physical intuition seems to show the same independence of rational or theoretical knowledge. Yet I respect it, and feel that it has an important place in any creative strategy. As an example, golden syrup is one of my favorite liquids (see chapter 8). My own physical intuition may, for example, have noticed its weird behavior when being poured. A thin stream of poured syrup sometimes forms a many-turned helix, whose lower coils disperse into the bulk fluid at the same rate as they accumulate at the top. Again, a thin string of poured syrup laid onto a polyethylene sheet, breaks up into droplets. The timing of the change, and the size and spacing of the droplets, says something about the surface tension between the sheet and the syrup. Those observations may be a snatch of physical intuition. Later, perhaps, a subset of them may add to others so as to feature in an experiment of mine or may provoke a "whim." But at present, they are just one of the many trivial things in my memory.

I understand neither human nor physical intuition, yet I like and value them both. Intuition, no matter how it arises, deserves some sort of mixed skepticism and respect. Thus Robert Townsend discusses the way a business manager develops a "gut feeling" that somebody is wrong for the job and sacks him.[3] Similarly, I knew a physicist who went into economics. He refused to take its theory seriously, essentially on intuitive grounds. "Real theory doesn't look like that," he said.

Another aspect of creativity is the steady following of a wrong train of thought. At the edge of knowledge, almost any argument can be proved wrong, or opposed by another of equal weight (that's why it *is* the edge). Despite this, a creative may pursue such an argument whether or not it makes sense. It seems interesting, or fun, or simply worth bearing in mind. My favorite real example is William Prout's nineteenth-century hypothesis that all atomic weights should be whole numbers. Accurate measurements disproved it (the atomic weight of natural chlorine, for example, is 35.453): yet it kept its chemical appeal. Thus it sparked Lord Rayleigh's discovery of argon. It is still enshrined in modern atomic theory, as the claim that every atomic nucleus contains a whole number of protons and a second whole number of neutrons. Daedalus, of course, followed up a lot of silly trains of thought. Some got published; some even came true! But most were pure nonsense. I now reckon that you should tolerate nonsense and even pursue it. The exercise may spark some notion you can use elsewhere: at the least, it may trigger a joke.

Lord Ernest Rutherford, the founder of nuclear physics and perhaps the greatest experimental physicist of the twentieth century, worked largely by his powerful intuition. His experiments were often very simple, but very insightful. An apocryphal story has it that a student of his once remarked, "The alpha-particles are his friends! He knows what they will do." Rutherford once phoned a student of his, Mark Oliphant, at 3 o'clock in the morning. Oliphant feared some bad news, but Rutherford said excitedly, "I've got it! Those short-range particles we saw are helium nuclei of mass three!" Still drugged by sleep, Oliphant asked him what possible reasons he could have for that conclusion. "Reasons? Reasons?" roared Rutherford. "I feel it in my water!" He was right, of course. The pair conducted a confirmatory experiment and later wrote for the scientific magazine *Nature* a calm and rational paper about their finding.

I reckon that intuitive notions inform and direct a lot of good practical science. And this leads me to expound some strange notions that Daedalus has played around with. Each relates to creativity and may well be untrue; but that has never stopped him. Even if it only has a small chance of furthering an argument, the occasional strange notion is worthwhile. So here is his scheme for enhancing creativity with a drug, something I despair of in chapter 3. He suggests that a woman can influence her baby by instructing her unconscious mind. And he proposes a gadget for recording dreams—thus perhaps spotting some of the creative ones I talk about in chapter 1.

No Sex in the Brain

Creativity may correlate rather with sexuality—so that it declines as age advances. I do not know whether the male sex hormone testosterone, or the female one estrogen, have any creative effects. But Daedalus has mused on the matter.[4] He has pointed out that sex hormones get into the brain at puberty. Initially, they are released into the blood. They then have to get through the blood-brain barrier. This keeps certain molecules out of the brain, while letting others in. One theory is that it consists of small holes, and works merely on molecular size (fig. 4.1). It admits small molecules like glucose and oxygen, which the brain needs, but excludes big ones, such as a sex hormone molecule bound to a large protein molecule. Testosterone itself is a small molecule and could get through the blood-brain barrier. But in males, it is mainly bound to one of two big protein molecules in the blood: albumin and "testosterone-binding globulin," or TeBG. In equilibrium, a small proportion of the hormone must escape from this binding. It gets into the brain, where it initiates typical male behavior and typical styles of male creativity.

In 1994 DREADCO's pharmacists invented Neutermind, a novel pharmaceutical that changed that equilibrium. The drug bound testosterone so firmly to its protein that it could not get into the brain at all. It could still go around the body, making the beard grow and sustaining other male physical features. But it was excluded from the brain!

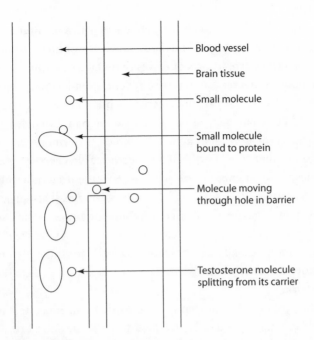

FIGURE 4.1   A Guess at the Blood-Brain Barrier

The blood vessel is on the left. It is separated from brain tissue on the right by a barrier that contains small holes. This lets in small molecules, such as glucose, oxygen, carbon dioxide, and little hormones. Large molecules, such as the protein molecule carrying a testosterone molecule, cannot get through. Only by splitting apart from its carrier can testosterone can pass through the hole to enter and influence the brain. Daedalus's drug Neutermind would prevent this splitting up from occurring.

DREADCO's Neutermind would be ideal for male soldiers required to be polite and courteous to colleagues of the opposite sex and male politicians required to maintain family values. It might damp the bitterness of the squabbles between male rivals in any creative field.

Daedalus saw another powerful use, too. From time to time, an artwork is sexually arousing. A critic may simply reject it as "pornographic." Now, at last, we have a simple test. Let the critic take a dose of Neutermind. He will cease to be bothered by any sexual implications. If the work is still interesting, it's art!

Testosterone has two recognized effects on the male brain. One promotes aggression, conquest, and dominance; the other promotes sex. These seem such different sorts of drives that Daedalus hoped to separate them. He planned to create two types of Neutermind. "Pure Power" Neutermind would eliminate sexual interest, while "Peace'n'Love" Neutermind would eliminate aggression. To create the pair, Daedalus espoused a shaky theory of hormonal action by which a hormone acts through its molecular vibrations. DREADCO's redoubtable chemists planned to change these by changing the atomic masses in the drug molecule.

Now creativity, like sexuality, declines with age. This suggests that testosterone may have a third effect on the male brain. It may stimulate the creative RIG. The "elixir of creativity" whose nonexistence I deplored in chapter 3 may actually be attainable. Sadly, Daedalus's plan for altering the atomic masses of testosterone (or its female equivalent, estrogen) will not work.

Yet his argument is not entirely absurd. I can imagine that some feasible drug might one day be invented to do some subset of whatever testosterone does to the brain. It might either enhance or dampen sexuality, enhance or dampen social dominance, or enhance or dampen creativity. Clever pharmacology might even invent variants for each function. At the moment, however, nobody can make any sort of Neutermind. My notion that creativity has some sort of correlation with sexuality is just a notion.

How to Have a Genius

A study of seven hundred eminent people has shown that most of them had troubled family backgrounds.[5] They were often trouble at school but had a lifelong interest in learning inspired by a scholastic home life dominated by their mother. I reckon that a woman has a subtle reproductive advantage. Even if she is not extremely creative herself, she may be able to have a child who is. Furthermore, she plays a major role in satisfying the curiosity of her offspring, answering its endless questions, and starting that store of knowledge. Daedalus points out that in any sex act, a man releases millions of sperms, but the woman only one egg or ovum. Furthermore, she releases a new egg only once a month. So sexual variety looks masculine. Not so! A woman is born with all her eggs already inside her—seven hundred thousand eggs per ovary or maybe one and a half million for the two. But at one ovum a month, only about four hundred ova are delivered in a woman's entire reproductive career. She must discard about four thousand eggs for every one she selects for fertilization.

How are the lucky ova selected? I suspect that the woman's unconscious mind influences the choosing. This is another good reason for trying to send instructions "downstairs" to it. Instructed or not, the unconscious mind of a woman will try to select eggs well matched to the men in her life and to the society that she sees around her. No matter what she claims for social consumption, or even believes herself with conscious sincerity, her unconscious mind may have its own ideas. It might think, for example: "This society rewards geniuses. So I'll have a genius!"

This theory explains a lot of human history. Some places have had large numbers of geniuses—I think of Ancient Athens, Renaissance Florence, Enlightenment Edinburgh—while others record none at all. Creative men may sprawl over the history books, but it was the women who set up the deals. Yet each child of genius may have made big trouble for its poor parents. Max Beerbohm feared that Milton, his linguistic paragon, must have been an awful child.

Awful children sometimes grow up into equally awful (or at least inadequately socialized) adults. We owe a collective debt of gratitude to innumerable unsung parents who allowed, or even encouraged, their awful offspring to grow up and do their thing. Ancient Athens, Renaissance Florence, and Enlightenment Edinburgh may have been hotbeds of unrecorded family trouble.

We owe a particular debt of gratitude to the mothers of those awful children. Far below conscious intention, the woman selects the few "good eggs" to be presented for fertilization. Daedalus noted that there is no connection, nerves or whatever, between the two ovaries of a woman.[6] Yet every month, one releases an ovum and the other holds back.

Do the ovaries take it in turns? Nature is seldom so democratic. But if the unconscious mind can "read" the eggs in an ovary, it will select and prepare its genetic favorites for the monthly honor. The left ovary, perhaps, goes in for order and predictability, and is favored in a fairly regular society. Conversely, the right one goes in for opportunism and flair, and comes to the fore under more chaotic social conditions. These two "rival colleges" compete by the sort of hormonal signals common in sexual physiology. Each ovary puts some hormonal compound into the bloodstream. In the circulation of the blood, this signal is read by the other ovary. Like two poker players, each raises the bidding until one folds in defeat. The other then sheds the ovum for that month.

So Daedalus proposed Anova, a subtle contraceptive pill. It would persuade each ovary that the other had won the monthly bid. Neither would then produce an ovum, and their owner would be infertile for that month. DREADCO endocrinologists were taking regular blood samples from female volunteers, looking for the hormonal signal put out by one ovary but read by the other. The "bidding" between them might be very simple—a single compound from one ovary saying, "I'm letting go this month" and discouraging the other from trying. Once the team had cracked the bidding code, Anova would be designed to subvert it. One pill once a month should do the trick.

Let's take the argument further. If the unconscious mind of the woman selects an ovum for release, maybe each ovum is equally strategic in its dealings with sperms. It may not just fuse with the first one it encounters; it may wait for a better match. Now there is a mathematical theory about selecting from a number of objects that are hard to change later. You can apply it to selecting a pub, or finding a house, or even choosing a marriage partner. You need to know what you want in a choice, the time required to make one, and the choices that are around. So you inspect and reject something like the first 1/eth of them ($e = 2.718 \ldots$), and settle for the next choice that is better than the ones you have rejected. An ovum, of course, may have millions of sperms to study, and not much time. It might be able to evaluate about ten sperms, which would mean rejecting (say) the first three and selecting from the next seven. An ovum is much bigger than a sperm. It has lots of room for strategic machinery. Launched into a mass of eager sperms, it will scan its suitors with a critical feminine eye, so to speak, seeking a good genetic match.

I do not know if ova studied under the microscope show any signs of this strategy. But I like the idea that we humans have invented a new evolutionary trick. After all, we evolved astonishingly rapidly—only about 5 million years from ape to modern man!

The male part of human reproduction is (as usual) to play the field. There are 22 human chromosomes, plus the XY pair that determines sex.

So a complete set of sperms, covering all possibilities, would consist of 2^{23} or 8 million sperms. This is not a bad estimate of the typical ejaculation of several hundred million sperms. Each sperm is little more than a mass of genetic material with a motor. It must "wear its heart on its sleeve" for ova to evaluate.

But why such a vast number? One theory invokes sperm competition. If a woman has several lovers, she is likely to become pregnant by the most prolific ejaculator. So as sperm are biologically quite cheap, it pays a man to deliver lots of them, so as to overwhelm possible competition.

How to Record Dreams

Your dreams may reveal a part of your unconscious mind. Many psychiatrists have tried to interpret dreams in this way (see chapter 1). Some people have found powerful new ideas during nocturnal dreams (see chapter 1); I have even had one myself (see chapter 11). There is no good theory of dreams, but if they say something about creativity, it is worth recording them—perhaps via the nerves.

A nerve works by a subtle ionic mechanism. Nerve signals in the body all seem to go in the direction the body needs; but a nerve impulse in the reverse direction looks entirely possible. This reminds Daedalus of dreams. We take in far more data that we can ever use; and one theory is that a dream is the brain's way of getting rid of trivia. It pushes unwanted data out by sending it backward, mainly to the eyes, which saw it in the first place.[7] This fits in neatly with the notion that rapid eye movement (REM) sleep coincides with dreaming. So why do dreams make no sense? Why are they not obviously a diary of rejected trivia? Ah, said Daedalus, like all brain data, they are not in the high-level language in which we consciously think. They are still rejects on the cutting-room floor of life and are still in the brain's internal machine code as they are pushed backward into the eyes.

If so, said Daedalus, a dream could be recorded. Any image on the retina, even from a dream, expands it slightly as the nerve's resting potential collapses. (Like any nerve, the retina reacts to pressure: hit in the eye we "see stars.") Daedalus set DREADCO biologists to design a set of goggles to capture that returning expanded image. Their idea was to detect small thickness changes in the retina by changes in its reflection of ultrasonic emissions. The goggles would be filled with simulated salty tear fluid, both to transmit the ultrasonic signal and its return, and to let a sleeping eye open briefly without smarting. When the goggles were perfected, DREADCO volunteers would wear them in bed.

Daedalus's ideas about dreams may not be entirely fanciful. Later, I came across an item in a 1948 issue of *Nature* that reported that the eyes of a living dytiscid beetle had been observed to emit light.[8] This major reversal of optics certainly supports Daedalus's claim that visual signals can go backward down an optic nerve.

Human ears often emit sound—typically a weak signal at a few hundred hertz. Doctors often test the ears of babies by trying to detect otoacoustic signals. Now if the eyes can receive dream signals coming the "wrong way" down a sensory nerve, so can the ears. Even better, the ear is a simple loudspeaker organ, much more easily reversed than the eye. I do not know if any physiologist has launched nerve pulses backward down an audio nerve and heard the ear speaking up in response. But since you can attach REM detectors around the eyes of a sleeper, you should be able to attach microphones to his or her ears. You could then, maybe, listen for the sounds of a dream.

This scheme looks very simple. I suppose the first step would be to try it on animals—I like the idea of attaching microphones to animal ears. What do animals dream about? Trivers might claim that they would dream about personal political information (see chapter 1). Like any human dream, this might be disguised in some way. Human interpreters should easily penetrate a merely animal disguise. If animal experiments gave interesting results, you could then try human subjects. Human data could test my bold claims that the human unconscious mind creates dreams, jokes, and new ideas.

5

Creativity in Scientific Papers

*P*robably every scientific achievement has a troubled human story behind it. The researcher is in it to achieve some RIG ambition, or one of his or her boss, or perhaps to sort out some puzzle. The final results, presented in a clear and believable publication, come much later. Most of the time, the worker is puzzled. Thus the great physicist John Wheeler said that the secret of effective research is to make the mistakes as fast as possible. Sometimes I waited years to reach a believable story. Two such cases are an interesting inversion of many of the stories in this book. The data were there for years. But the RIG ideas that made sense of them came very late. I do not regret that wait. A good scientist is not just in the business of presenting results; he should believe his story.

A Chemical "Garden" in Space

In 1988 I met Ulrich Walter, an astronaut for the German Space Agency. He invited me to suggest a "pocket experiment" for the D2 mission—the second German launch of the Space Shuttle, then targeted for 1991. I agreed at once. I recalled going to the chemistry department at University of California, Berkeley, where George Pimentel showed me their huge machine shop. "This is where we built the spectrometer that went to Mars," he said. Wow! What a lovely boast! I began to dream up space experiments.

The absence of gravity seemed the central challenge. What would happen, for example, to a bubble growing electrochemically on an electrode? I built an apparatus for Ulrich to try on the "vomit comet" (see chapter 16) but got ambiguous results. A second idea of mine was to blow a soap bubble in zero gravity. I imagined an apparatus to do it—and later worked out what it would do (or so I thought).

The D2 committee for space experiments accepted yet another idea. It was for a chemical garden in space. If you drop a crystal of a metal salt (such as cobalt chloride) into sodium silicate solution, a sort of "tree" of insoluble metal silicate grows up from it, against gravity (fig. 5.1). If there are several different metal salts, a "garden" of plantlike objects forms. What shape would such a garden take in zero gravity? I played with this reaction and failed to make any firm prediction. So it seemed a good space experiment to try.

I began to imagine an apparatus to grow a chemical garden in the Shuttle. I still have many of my designs—the naivety of the earlier ones shows how far my ideas advanced as I developed the equipment. But several of my early decisions turned out to be vital. First, we would build three units. Two would fly in space, and I would work one on the ground, as a sort of control. Each unit would have two injectors in it, so that it could grow two chemical gardens; if the gardens in a space unit met, the collision could be interesting. We also agreed that Ulrich would record the gardens photographically—the electronic video download of the D2 mission was saturated already.

So I began to look at the chemistry. I played with a vast number of metal salts and sodium silicate solutions, seeking the fastest-growing and most interesting chemical gardens. I also played with crystals that had been dusted with magnetic powders, so that I could divert chemical gardens from their vertical growth with a magnet. My good friend Fred Peacock, who had been a student with me at Imperial College, was now head of chemical research at Berol (a pen company). He sent me lots of dyes. In the course of all this chemical play, I invented a strong solution of a metal salt, thickened to a paste with fumed silica (a light dust that had long been one of my favorite solids).

I can only thank the physical intuition of my unconscious mind for such a brilliant invention. That paste did not emerge from any sudden burst of enlightenment but just came out of my chemical play. In normal gravity, it grew an interesting chemical garden and provided a totally new way of growing one in space—quite a different challenge from the one I had been wrestling with, that of injecting a solid crystal into a silicate

FIGURE 5.1 A Chemical Garden in Normal Gravity
If you drop a crystal of a metal salt (such as cobalt chloride, shown here) into
sodium silicate solution, a sort of "tree" of insoluble metal silicate grows up
from it against gravity. If there are several different metal salts, a "garden" of
plantlike objects forms.

solution. Colored with one of Fred's dyes, a paste in zero gravity might release streams of dye and show something about how the liquid flowed around it.

I had one hour of astronaut time. This brevity worried me. Could my reaction, slowed by the absence of gravity, get anywhere in an hour? I chose my metals and silicate concentrations to speed up the chemistry as much as I could. Fortunately, once in space Ulrich found time to take more photographs after 48 hours. Further developments were obvious.

It took me many months to design the units. The final design is shown in figure 5.2. Each chemical garden unit had two different injectors. One punched a single crystal on a rod through a thin curved membrane into the silicate solution. The other injected that clever dyed paste of metal salt solution. Crystal or paste, one should show some interesting zero-gravity chemistry.

I was later very pleased about my design for a lip on the paste injector. My idea was that if the space garden happened to hit it, it would bounce off or show some informative interaction. Amazingly, one paste garden did just that (though not in the way I had imagined). Another part of the experiment was quite unintended—a fiber from my woolly pullover attached itself to the sealing grease of the ball closing the paste pipe. In the hurry of assembly, there was no time to do anything. That fiber would just have to be part of the experiment. In the event, it was mightily useful—the garden just grew through it. That showed that a garden forming in zero gravity did so as a fluid.

This was sheer dumb luck. But my whole apparatus was riddled with tricks that might have given information, though Mother Nature (as usual) was cleverer than I was. Thus on each unit, the crystal for injection was a single crystal. All its molecules were aligned in the same lattice direction. If a chemical garden grown in space had shown any directionality, I should have spotted it. In the event, I saw nothing.

I did not build the units myself. Two brilliant Newcastle chemistry department engineers, Roland Graham and Bruce Atkinson, did the hard work. They would make anything to my design, or criticize or veto it, or say that an easier solution was possible. Indeed, I often "subcontracted" elements of design to their expertise. Most of the O-ring pressure seals for the units were theirs, for example, and Roland made the "moat" rubber seals for the outer shield of each unit—I still do not know how. But I de-

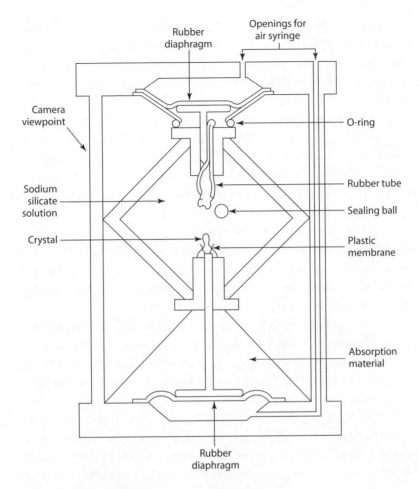

FIGURE 5.2   A Chemical Garden Unit Used in Space

The top and bottom ends of the unit are opaque metal. When called upon in space, it is worked by air forced from a syringe into openings. These drive the rubber diaphragms, shown in their halfway positions. The lower diaphragm pushes a crystal through a thin plastic membrane into sodium silicate solution. The upper diaphragm pushes a plunger whose compression squeezes a rubber tube and ejects its sealing ball. The tube contains a toothpaste-like suspension of metal-salt solution, which it extrudes into the sodium silicate solution. That solution, and the cell containing it and the sides of the apparatus, are transparent. A camera looking in takes photographs and videos of the zero-gravity chemical gardens as they develop. The lower region contains absorption material to mop up any solution that might leak from the cell. A rubber O-ring stabilizes the cell against the vibration of the rocket launch.

fined the design and layout of the units and did some of the constructional work on my old lathe. Cleverly, I designed each unit so that a camera could rotate around it on one surrounding ball-race axis, and would give a true x, y, z view into the cubic cell. That photographic mount would take both video and then-standard 35mm still cameras, as carried on the D2 mission. To my engineers, the mount was unfeasibly flimsy—I told them that there was no weight in space. Further, I had a lot of trouble making the cubic cell for each unit. NASA insisted on polycarbonate (as used in bulletproof windshields). Only at a late stage, did I discover a splendid glue for polycarbonate. Had I known of that glue earlier, I could have made the design better. Oh well.

In organization terms, I felt like a sort of pig-in-the-middle. Above me was Klaus Kramp of the German D2 team and above him NASA. Both wanted a firm design for the apparatus they were going to put into the Shuttle. Below me were the engineers building the units. And I annoyed my superiors by continuing to change the design.

However, those superiors were ultimately on my side. At one stage, the German Space Agency wanted delivery so soon that I simply gave up. I told Klaus Kramp to cancel the experiment. He then told me that many other experiments were behind schedule as well, as was the launch of the mission itself. He also helped me out when the completed experiment, two units and their accessories in a box, had to be submitted to a vibration test. We might have done this in Newcastle, but Kramp saw the chance to use the proper space-vibration equipment in Holland. We struggled to get the box completed for this test. Of all the time pressures we fought against, it was perhaps the most pressing; for I was still designing the units. And in that panic, I encountered what was for me a unique mental event. The heavy reaction cube, full of sodium silicate solution, was retained by only one nut at one end. Vibration could shake it loose. Roland, Bruce, and myself all worried about the problem. Then we saw how a simple silicone-rubber O-ring could solve it. It needed a bit of elementary machining, but it worked. It remains the only example I know of collective creativity!

I had a lot of trouble with the NASA bureaucracy. It seemed not to like a nasty chemical experiment flying in its nice Shuttle. I gave NASA a vast amount of toxicological data on the chemicals I planned to fly. Even so, NASA insisted on at least two layers of containment, each to withstand several atmospheres of pressure.

My main chemical worry was something NASA never asked about. My apparatus was made of a light alloy, which was mainly aluminum. Its surface would slowly and slightly react with sodium silicate solution, giving a little hydrogen. This was a shocking state of affairs for a space experiment. As Roland and Bruce made each light-alloy component, I boiled it in silicate solution until no more hydrogen was given off. My idea was that I had then "immunized" that item against sodium silicate, probably by covering it with a protective coating of silica. If all went well, of course, the silicate solution would never get out of its cell and would never touch the metal.

NASA seemed unconcerned with whether the experiment would work: that was my concern. But it was adamant that it be safe. Thus in each unit we made room for an "absorption material" that could mop up all the sodium silicate solution if it leaked out. I tried a number of absorption materials. The best was ARCO's Fibersorb, which the company had made for bandages and disposable diapers but had not commercialized. An absorbent polymer, FSA from Allied Colloids, was also very good. Such materials could absorb hundreds of times their weight of water. Each unit held its absorption material in a stainless-steel wire-mesh container. I liked that fiber, for it could not get out. A pure absorptive powder, such as Dow's Drytech, might in zero gravity have spread about inside the unit and obscured the optical view. In any event, we had no solution leakages and the absorption material was never called upon.

Another aspect of that design is relevant to this book. All through the construction, I had been running away from one problem that I had to solve. How to start the experiment going? One of my biggest fears was that when Ulrich in the Shuttle pressed the button to work the thing, nothing would happen. I knew that my apparatus would be stored for months, during which time no reaction at all must occur. One solution was to trigger the apparatus pneumatically, with air pressure. But I did not like air pressure—I suppose my RIG did not trust it. It preferred a firm, reliable mechanical system. Indeed, I once felt I had invented one and began to build it. But time and NASA pressed, and I ultimately went pneumatic. A splendid northern UK firm made flexible diaphragms to my design, and I worked endlessly to reduce the pressure needed to drive them and work the injectors from them.

Finally, the chemical garden box was installed in the Shuttle. When NASA took reference photographs of its two units, the pictures showed that one of my paste injectors was clearly leaking! I was worried. My Observer-Reasoner devised a clever argument that it did not matter much. But of course it did. My silicone rubber tube was a feeble osmotic membrane; but it had months of storage in which to act. I fought the notion internally but could not dodge the duty of testing it in the laboratory. As I feared, sodium silicate solution was slowly being taken up into the silicone rubber tube by osmotic action and was pushing the paste out. Hence the slow leak.

Later the Shuttle, with 3 seconds to go, was halted in its launch. As a result, the whole lift-off was delayed by several weeks. Many experiments, such as the biological ones, had to be re-prepared and re-packed. Klaus Kramp knew that we had made three units, of which only two had to fly. Brilliantly, he suggested that I should prepare the third unit and put it in the Shuttle to replace the leaky one. His wonderful organization made this feasible, too! So we did it. In my replacement unit, I replaced the silicone rubber tubing with butyl rubber, which is far less osmotic. The best I could find in the time was valve rubber tubing for a bicycle tire. This hasty lash-up worked. In due course I got the leaky unit back and found to my horror that both injectors had moved. It seems that the filled Shuttle had been pumped up a bit with air, to see if it was leaking. Nobody had thought to tell the experimenters.

So my RIG had been right to shy away from pneumatics! Furthermore, in working with pneumatics I had been far too clever. My pneumatic system was sensitive even to high atmospheric pressure. I worried endlessly about this. What was the air pressure in the workshop when the units were sealed? What was the air pressure in Florida, from where the Shuttle was to be launched? The meteorologists could not reassure me. Of course, I had given the units seals against the changes in air pressure caused by air transport. In retrospect I could have avoided all the worry about pressure by specifying that these seals should only be released in orbit. But I didn't think of it. I didn't even know it could be done.

So I just worried. I recalled a line from a film script uttered by a character played by John Cleese: "It's not the despair. I can stand the despair. It's the hope!" Yet with the equipment out of my hands, I had to fret. Mercifully the air in the Shuttle was kept at constant pressure even during launch, and my pneumatic apparatus worked. Three out of the four gardens grew properly when Ulrich operated them; one crystal injector failed.

That failure was bad chemical luck. The adhesive I had devised to glue the crystal weakened during months of storage. But I had several equally undeserved strokes of good luck. The most important one was that the space gardens survived reentry and even transport to my laboratory in Newcastle! I was able to study them under a microscope. Another happened a year later when, still not understanding my results, I went to a mission conference on the German island of Norderney. By some bureaucratic muddle, I was not given a file of conference papers. So I listened to the lecturers. Dieter Langbein's initial comments were not in his filed paper. He talked about the way two liquids can behave in a cubical cell in zero gravity. One liquid can form a thin film on the walls of the cell.

Later still, I was able to put the returned space-grown gardens in a scanning electron microscope. I had never expected to get them back, and my units could not easily come apart. I had to invent a way of opening them and of drying and mounting the space gardens for microscopy. The calcium garden, which from the space photographs I thought had failed to grow, had done something most surprising. It had formed a silicate that stayed liquid for hours. In zero gravity this had coated the walls of the cell as a thin film, which later solidified. I totally failed to spot the film. I even took photographs through it, pictures of the other garden in the cell. Luckily, when I drained each unit to extract its space gardens for microscopy, I spotted that film. I recalled Langbein's lecture and stained the cell wall with one of Fred Peacock's dyes, Arianor Mahogany. The thin calcium silicate film showed up clearly.

Then I had to dry the wet chemical space gardens for the electron microscope. Annoyingly, a dried chemical garden just breaks up. But there's a trick sometimes used to dry biological specimens. At 25°C, carbon dioxide gas can be compressed to a liquid miscible with water. You can flush a frail wet solid with it, displacing the water and wetting it with the liquid. Then you warm the specimen to 35°C—which is still below body heat. Without any boiling or violence, the liquid carbon dioxide becomes a compressed gas. You can leak it away, to get a fine dry solid specimen. An expert in a local hospital research team pitched in and did the drying.

Even when I had lots of results (including that unexpected but welcome electron-microscope data), it took me years to understand the experiment. I slogged through many mathematically dense physics papers on fluid instabilities. Finally, in a paper on quite a different matter (M. Glicksman's work on the solidification of succinonitrile), I saw one of his photographs. Aha! I thought, "That's what my space experiment did!" The key (which my unconscious mind had never imagined when I designed the units) was that in zero gravity, in the absence of all convection, the reaction gives a liquid. This later solidifies, but in the form of the liquid. My unconscious mind was convinced. I "bought" this explanation and wrote a paper about it.[1] Later I wrote a less formal account of the work.[2] Later still I looked at the statistics of my results. The gardens had done two things I had predicted. They had failed to do two things I had predicted. They had done six things I did not predict. My score was thus 20% (see chapter 1). That's about right even for serious science! And looking now at the weight of my designs, and the correspondence I had about them with the space authorities, I salute the force of the well-known aerospace industry saying, "When the weight of paper equals the weight of metal, it will fly!"

Incidentally, when the experiment had been successfully completed, and I was returning to Newcastle from the D2 control center, I realized at the airport that I was going down with a mighty cold. I was stricken with that illness for about a month afterward; and I now reckon the events were related. I had been keyed up for almost a year building that experiment; when it worked my bodily system just let go. Mind does indeed affect body!

The whole story shows the interaction of my unconscious mind and my Observer-Reasoner. NASA had specified the size and weight of the reaction unit; and my design met it. My RIG had given it two quite different injectors for getting the reagents together; both contributed to the findings. I knew almost nothing about what would happen. But I exploited good luck when I got it and wrestled with bad luck. I waited for years to publish the results, even saying nothing at the Nordeney conference. Only when I felt I understood what had happened, that is, when my unconscious mind was happy with my claims, did I go ahead. The final results agreed with my unconscious intuitions, and with the logic of my Observer-Reasoner.

A Past Puzzle: Some Funny Chemistry

This is an account of some chemistry experiments I did decades ago. Some of the results puzzled me mightily. I took the coward's way out—I left them out of the papers I wrote. I came to disbelieve a claim that I could have easily made and published and that the scientific world would have accepted, at least until a better chemist studied the matter. But I held back until I had a story I believed. I did not reach the full truth—but I made a step on the road.

I was studying the compounds formed by adding aluminum chloride to various simple organic substances. The molecules just add, with very little change on either side. The point of the project was to make aluminum chloride complexes with as many different organics as I could and to rank them for chemical stability. I hoped to see how strong the chemical bond was between the aluminum chloride and the organic molecule. I could then compare the chemical strength of the bond with its physical strength, as measured by far infrared spectroscopy. Spectroscopic data and complex calculations might let me extract the physical strength, the "force constant" of the bond. There was no compelling theoretical reason to expect much of a ranking, but a physical bond strength might be a useful guide to the chemical one.

Aluminum chloride is tricky to handle. It and its compounds react readily with the water vapor in the air. Indeed, in the absence of other molecules, it reacts with itself.[3] Thus the vapor is mainly $(AlCl_3)_2$, with a little $(AlCl_3)_3$; an elevation-of-boiling-point measurement I once made on a solution of it gave $(AlCl_3)_4$. I got my aluminum chloride in big bottles containing a somewhat impure solid. My first task was to purify it, which I did by heating it with dry salt in a special apparatus. This mixture evolved a vapor of pure aluminum chloride, which on cooling condensed directly to the solid. I collected it in a glass tube and transferred it to a special bottle in a "dry box," whose air was kept carefully dry.

The substances I planned to react with my aluminum chloride had to be pure and dry as well. They were less trouble to handle, for they did not react with the water vapor in the air. They were mainly liquids. From each I removed water with phosphorus pentoxide (which reacts with even a trace of water). I decanted off the liquid and distilled it in dry air to get a dry material. All this is part of the chemical art. One of the caveats you learn as a chemical researcher is "never believe the label on the bottle!"— always subject the material to your own purification routine. It even helps to check that the stuff is indeed what it claims to be.

Having got the materials pure and dry, the next problem was to make them react together. I made a number of compounds, dissolved them in suitable solvents, and managed to obtain believable far infrared spectra (at the time that was pushing the state of the art). One of the things I tried was reacting my aluminum chloride with the liquid acetonitrile. Chemists know acetonitrile as CH_3CN because the two carbon atoms are very different. And they aim for a product with good crystals of sharp melting-point. Both imply purity.

Aluminum chloride dissolves well in acetonitrile. The solution gives sharp-melting crystals. With acetonitrile as X, I made them $AlCl_3,2X$. The obvious interpretation (made by me and by the chemists who had made the stuff before me) was that it was composed of $X-AlCl_3-X$ molecules. But its crystals did not seem like those of the other complexes that I had made. They were feathery and flocculent, even maybe a bit greasy: not the solid, gritty crystals I was used to. And as a good experimental chemist, I was sensitive to the touch and feel of crystals. Worse, the new compound did not dissolve in solvents such as benzene and chloroform that dissolved other substances in my study. In fact I could only dissolve it in more acetonitrile.

Even stranger was its behavior under vacuum. The solid was very stable. But if heated to the melting point, it gave off lots of acetonitrile vapor: indeed, it became $AlCl_3,1½X$. The crystals of this compound were feathery and flocculent, rather like those of another strange compound I had made, $AlCl_3,1½Y$, where Y was methyl formate. I invented clever formulae for my strange 1:2 and 1:1½ compounds (fig. 5.3 a and b). But 1:1½ compounds are impossible as such. Half a molecule cannot exist. My formula (b) would have taken it as 2:3. The doublet $(AlCl_3)_2$ certainly exists in the vapor, so my argument was not chemically absurd. But my unconscious mind could not believe it. I wrote a couple of papers about my research but left the troublesome compounds out of them.[4]

Later, in the midst of other work, I kept looking at those products. I noticed (what I should have seen years before) that the infrared spectrum

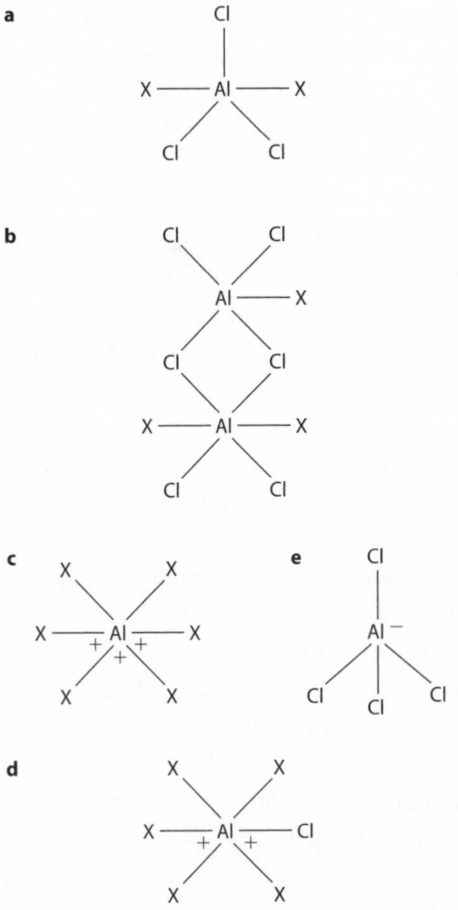

a

b

c

e

d

FIGURE 5.3   Molecular Structures for Aluminum Chloride Compounds

I once combined aluminum chloride ($AlCl_3$) with acetonitrile (CH_3CN, which I called X). Like chemists before me, I got $AlCl_3.2X$ (*a*), implying a ratio of 1:2 between the ingredients. I could also make another compound, which had the enigmatic ratio 1:1½. I guessed it was $(AlCl_3)_2.3X$ (*b*), with a ratio of 2:3. $AlCl_3$ can form a double molecule, so this was not chemically absurd. But I neither believed nor published it.

Years later I suddenly had quite a different idea, which I feel was a prompting from my RIG. Aha! I thought. My 1:1½ compound was the basic one. It formed the misleading 1:2 compound $AlCl_3.2X$ by adding "acetonitrile of crystallization." It was an assembly of charged molecules (ions), (*c*) and (*e*). This fitted the facts, and I published happily. Later I found that another chemist had preferred *d* to *e*. Indeed, the reaction may have many products. I am pleased that I held back.

~~~~~ A Few Words about Chemical Notation ~~~~~

Everything is made of molecules. The chemist shows the "molecular structure" of a compound by giving its constituent atoms a one- or two-letter abbreviation: aluminum = Al, chlorine = Cl, carbon = C, and so on. A molecule with several atoms has a subscript. Thus aluminum chloride is $AlCl_3$—it has three chlorine atoms. A double molecule is shown with a parenthesized subscript, as in $(AlCl_3)_2$. An unknown or abbreviated atom or molecule may be X or Y. A plurality of molecules takes a front number (as in 2X). A chemical bond can take a period; a looser one, a comma. A molecular structure is a sort of map of atoms, with the chemical bonds between them shown as straight lines. Some substances do not have a defined molecular structure. They are an assembly of charged molecules (ions), such as $AlCl_4^-$ shown as (*e*) above. Ionic compounds are quite well known. That charge (plus or minus) is superscripted. In any real ionic compound, the charges add to zero.

of $AlCl_3,2X$ was much like that of $AlCl_3,1\frac{1}{2}X$. This indicated that they were much the same—though I was still thinking of them as entirely different. And both spectra seemed to show the infrared bands of the ion (charged molecule) $AlCl_4^-$, which I happened to know well.

In a sudden aha! moment, I saw that my basic compound did indeed have that enigmatic 1:1½ ratio. Acetonitrile is a liquid, like water. Added "water of crystallization" is well known. Maybe enough "acetonitrile of crystallization" could make those misleading crystals into $AlCl_3,2X$? When I destroyed those crystals by melting them under vacuum, I could pump away excess acetonitrile as the vapor. This left $AlCl3,1\frac{1}{2}X$, just like another strange compound of mine, $AlCl_3,1\frac{1}{2}Y$ (methyl formate). That prompting from my RIG swept away guesses *a* and *b* above. The new compounds were ionic clusters! That was why their crystals felt strange to my spatula and why they did not dissolve in the usual solvents. With $CH_3CN = X$, I imagined that $AlCl_3,1\frac{1}{2}X$ was $AlX_6^{+++},3AlCl4^-$, *c* plus three of *e*. The whole assembly then added two molecules of "acetonitrile of crystallization," giving four $AlCl_3.2X$ units. It all made sense!

My unconscious mind was convinced; I felt happy with my claims. My publication was accepted.[5] Much later I came across a related paper.[6] It also felt that $AlCl_3.2CH_3CN$ contains "acetonitrile of crystallization" and that the basic product is ionic. In fact, there may be many products; my crystals may have grown just because they were relatively insoluble. I am still glad that I never published about *a* or *b*.

The Stability of the Bicycle

When I was young, I discovered that I could ride a bicycle when I was too drunk to walk. Bicycle stability is clearly a problem in physics.

Years later I started playing with bicycles again. I soon disbelieved the conventional "encyclopedia" theory, in which the front wheel acts as a gyroscope. (I suspect that this idea is based on the Brennan monorail vehicle of 1912, which was indeed gyro-stabilized.) Having decided that a gyroscope was absurd, I proposed it in a Daedalus column.[7] R. Hobart Ellis, the editor of *Physics Today*, invited me to write a paper on the problem for his journal. It became one of my most important papers.[8] As a chemist

I felt free to be a bit informal in a physics journal. My light-hearted style appealed to the readers. It may be the only research paper ever reprinted![9]

To try some experiments, I bought two secondhand bicycles to vandalize. First I explored the conventional gyro theory, for which I built URB1 (Unrideable Bicycle 1). I mounted an additional wheel on its front forks, as a gyro or anti-gyro element. I mused about, but rejected, the notion of gearing the two together. The engineering would have been tricky. Instead, I mounted the axle of the extra wheel on the front fork of URB1. The new wheel did not touch the ground and could be spun freely either way. It held that spin for ages. I could ride URB1 easily, no matter which way the gyro wheel was spinning. Crucially, I could ride it with no hands, even with the extra wheel spinning the "wrong" way. The bicycle was indeed helping its rider, and the gyroscopic theory had to be wrong.

My second experimental bicycle, URB2, had a very tiny front wheel: in fact it was a furniture castor some 2 centimeters in diameter. I fastened it to the bottom of the front wheel of my test bicycle and locked the front brake with a clamp. Only the little front wheel could spin. I could ride URB2 as well but learned little from it. The castor got very hot; furthermore it could not go over a bump more than 1 centimeter high.

My third bicycle tested a notion that came to me while I was riding URB1. That bicycle was a bit awkward to ride, a fact I associated with the high moment of inertia about its steering axis. The steering forks had not only to turn the front wheel, but the extra gyro wheel as well.

So I studied the steering axis inertia of my main test bicycle. Its front basket swung with the steering forks, and I put concrete blocks in it. I then pushed the bicycle away and watched its motion. Then I reduced that moment of inertia as much as possible, by removing the front basket, the front handlebars and the front brake assembly. Idly I turned the front wheel all the way around (you can't do this on this all bicycles, but on this stripped-down one I could, luckily) and pushed the bicycle away. It stayed up! It kept going! It even corrected its initial wobbles! Quite by accident, I had made an astonishingly stable bicycle (fig. 5.4). This was a real aha! moment. I did this experiment in the ICI parking lot. The company parking lot attendant tried to stop me; but I told him I was a member of staff and kept going.

I was amazed. Later, I saw this experience as a lovely example of the creative process I am advocating.

1. I was doing something quite new—I was pushing a test bicycle with concrete blocks in the front basket but not getting on it.
2. I then removed the whole assembly, for a good scientific reason. This turned out significant for quite another reason.
3. My apparatus was so simple that I could indeed play with it—I could turn the front wheel of my test bicycle all the way around.
4. I then tried an idle experiment, with no rationale at all but
5. I got an amazing and unexpected result and then
6. I spotted its significance.
7. This pushed the whole research in a new direction and gave me a new understanding.
8. To do the experiment at all, I had to overcome authority, in the form of the ICI attendant; this I was fortunately able to do.

I called my new bicycle URB3. I reckoned that it owed its stability to its unusual front-wheel steering geometry. I began to explore the problem on the ICI computer, an old-style IBM monster (in those days computers were big, special beasts). The mathematics were very tricky, and I soon gave up a direct approach. Instead I lashed up a subroutine with an iterative step in it. I kept running it until the errors were very small. (Later a French correspondent, E. Soulié of Chatenay-Malabry, did the job properly. He presented me with a routine that generated the angles and dimensions directly in one step.) URB3, though amazingly stable, was not easy for me to ride. I suspect that I kept trying to steer it. A drugged rider, happy to go anywhere, might do better.

Gyroscopic theory says nothing about steering geometry. It does not explain why all front forks bend, and no steering axis is 90°. By contrast, my theory emphasizes it. A bicycle needs its front wheel to touch the ground a little behind the steering axis. (Note that the T in figure 5.4 is small.) With a large T (T' in fig. 5.4), the bicycle is like URB3, too stable to be controllable. Racing bicycles, for which maneuverability is more important than stability, have a very small T. All this is already known in practice to bicycle designers. My work merely brings physics up to where engineering has already gone.

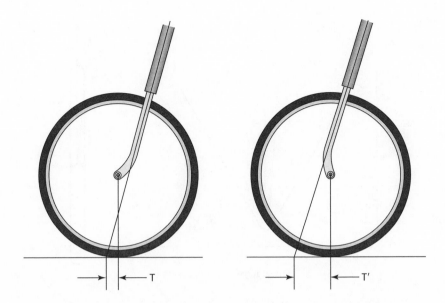

FIGURE 5.4   Front Bicycle Wheels
A bicycle is stable and rideable with T small and positive. The usual front-wheel geometry (as on the left) ensures this. A bicycle such as my Unrideable Bicycle 3 (URB3) has its wheel reversed, as on the right. This makes T' large and positive. The bicycle is too stable to ride.

I made URB4 simply by pushing the front wheel of my test bicycle about 10 centimeters forward. I simply bolted commercial angle plates to its front forks and mounted the front wheel on them. This shifted its steering geometry well into instability. URB4 was gratifyingly unstable, though I could still manage to ride it. It had no inherent stability and crashed to the ground when pushed and released.

That bicycle paper annoyed my research boss at ICI. Luckily Duncan Davies, head of all ICI research, was greatly amused—it showed the wide range of thought of his researchers. Furthermore, I am now hailed as the father of modern bicycle theory.[10] Much later my friend David Taylor made a series of films about the bicycle. I contributed to the filming, part of which used the 500 meter runway of Brunton Park airfield. That runway had a slight downhill slope, and my stable bicycle ran down its whole length with nobody on it!

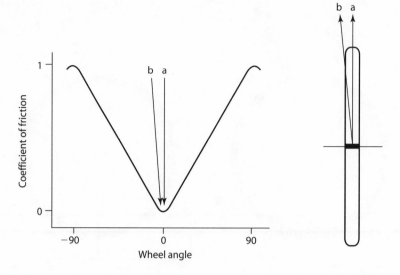

FIGURE 5.5 How a Wheel Works

A wheel rolls best at right angles to its own axle, as shown on the right by line *a*. All steering-gear systems use this fact; a bicycle or a car travels in a direction defined and controlled by the angle of its front wheel(s). If forced to roll out of its natural line, for example along line *b*, a wheel rolls with increased friction. The wheel of a mathematical planimeter may be forced to roll up to 90° away from its "natural" direction of roll. Its coefficient of friction— the force by which the wheel resists movement, divided by the force being applied to it—increases markedly with that deviation, as shown in the graph on the left. I was interested in what happens at very small deviations from a wheel's natural direction of roll. I designed a "tricycle" vehicle to explore the matter. It would have had two wheels on ball races, intended to run freely and fixedly in plane, and a third wheel whose deviation could be varied slightly. In the event, I never built the vehicle and so never reported any findings. The whole scheme remains one of my many unrealized fantasies.

I had in mind to write another paper on this topic. It would have been called simply "How does a wheel work?" All car steering depends on the fact that a wheel rolls most easily "in plane," at right angles to its axle. Pushed parallel to its axle, a wheel does not turn at all, and pushed at a lesser angle it turns stiffly and slowly at the speed appropriate to its in-plane direction. Indeed, this is how a mathematical planimeter works. I liked the idea of studying that portion of the wheel curve very slightly out of true. It may be well known to engineers, but not to physicists.

My plan was to build a tricycle-vehicle (fig. 5.5). Like any tripod, a tricycle has a calculable load on each wheel, even if its "road" is not perfect. A tire on each wheel would give it some known frictional character. Two of the wheels would be accurately in plane, perhaps running on ball races. The third wheel would be adjustable, maybe by some threaded adjuster that could make it as true as the others or out of true by a known amount. Each wheel would leave a track. I hoped to look carefully at that small section of the wheel curve around a to b in the diagram. I wanted to measure the force needed to pull the vehicle at several speeds, while studying the effect of the slightly wrong wheel. Alas, I never did the work or wrote the paper and have no idea what I might have found.

6

Heat and Gravity

The mind of a creative person is never idle. Consider James Prescott Joule on his Swiss honeymoon in 1847. His young bride did not distract him from scientific thought. Indeed, he took a sensitive thermometer with them. And whenever the couple came across a river waterfall, he measured the temperature of the water at the top and at the bottom. The mechanical equivalent of heat was brewing in his mind, and waterfalls could be part of the evidence. On reasonable assumptions about the flow, the water should get hotter by 0.00239 degrees Celsius for every meter it fell. The rivers seemed to support his theory. He ultimately determined that equivalent, took the idea further, and the unit of energy is named after him. Whatever her private misgivings, his new wife went along with her scientific husband. I know the feeling Joule must have experienced—a thinker is never off duty. In this chapter I recount some of my own encounters with heat and gravity.

Must Heat Rise?

Heat rises. This unavoidable truth governs all fluid flow. It drives meteorology and oceanography and bedevils the reaching a comfortable temperature in our own homes—as hot air goes up and mainly heats the ceilings. For like almost everything, hot gas expands on heating and rises. Is there a way out?

I recall from my schooldays the strange reversible reaction between the gases N_2O_4 and NO_2:

$$N_2O_4 = 2NO_2$$

This is an equilibrium. It can go either way. In this case, heat drives the reaction to the right, as a chemist would expect. A big gaseous molecule breaks up into two small ones. On heating, gaseous N_2O_4 gets lighter than cold gas even faster than usual. It is even more convective than air. Might there be a reaction that went the other way? If so, you might make air nonconvective, and stop heat rising! My RIG liked the idea of nonconvective air so much that Daedalus claimed (falsely, I fear) that you can make it nonconvective by putting a little methyl formate in it.[1]

Despite this absurdity, you can indeed muck about with density to stop heat rising. One such trick is the "solar pond." Sunlight shines into a big shallow lake. Light goes through clear water with very little loss, so the brownish bottom of the lake absorbs most of the sunlight. Then you throw salt in. The salt sinks to the bottom and dissolves there. The resulting solution has a concentration-gradient: strong at the bottom (where the solution has a high density) and weak at the top. With good design, this stratified liquid does not convect. If you don't stir the pond, the gradient can last for years. So the energy of the sunlight absorbed by the bottom just stays there. Static salt water is a very bad conductor of heat, so a high bottom temperature, approaching 100°C, can build up and pipework can extract the heat.

Daedalus once adopted the solar-pond notion to work with heavy gases. He planned to use it to generate solar power.[2] And I sometimes play with gas-densities high and low, quite without asking what use the results will be. Thus while playing with amateur balloons, I discovered that the air in a room is highly sensitive to very small changes of temperature. It slowly circulates thermally, rising above a heater and sinking next to a cold wall.

My balloons were plastic bags and condoms, filled with domestic gas (which is methane, and buoyant in air). I tied each one with a string and weighted it to be about neutrally buoyant. Each balloon then floated stably in the air. In a room with a fire, my balloons followed the slow thermal circulation of the air. Near the fire, they rose slowly toward the ceiling but just failed to touch it. They drifted across the ceiling and down the cold opposite wall but just failed to ground on the floor. Instead, they followed the slow room-air circulation back to the source of heat. Yorkshire Television Ltd. filmed a polished version of my domestic experiment. We didn't use bags or condoms; nor did we fill our balloons with methane gas. We used Mylar balloons, weighted with a sticky flexible plastic called Blu-Tack and filled with air and helium.

So a crazy experiment turned into something to be broadcast on television for a broad audience. But I continued to muse on domestic heating. My scheme to raise its efficiency with argon (see chapter 15) mentions its effect on small ground-based pet animals. This little remark derives from my own experience.

When my brother's female cat had kittens, I built a maze for them out of Styrofoam ceiling tiles. Astonishingly, they made it their home! Later I realized why. Styrofoam is a splendid heat insulator. If you are a small animal on the floor, you are very sensitive to cold floor-based drafts. Hot argon-loaded air that did not convect up to the ceiling, but stayed on the floor, would be welcomed by pets as well as by human beings!

And I continue to muse on gas density. The densest gas I could get hold of was Arcton 114, an ICI refrigerant (dichlortetrafluorethane). It is about six times denser than air and quite invisible. In a rubber balloon, it makes a crazy "lead balloon" that falls like a stone. You can even fill an aquarium tank with it (the one I used was about 30 centimeters across and 70 centimeters long), where it stays for hours. You can then float a boat on it. The boat appears to float on nothing, for no sharp edge is ever visible. That experiment may have sensitized my RIG to the significance of gas density. If so, it may have sparked some of the ideas in this chapter.

The Artificial Geyser

The geyser is a natural geothermal phenomenon. At regular intervals, a jet of boiling water spouts out of a hole in the ground, often at the bottom of a small lake. In between eruptions the geyser is essentially quiescent. Nearly all geysers are in New Zealand, Iceland, and the United States. They depend on some geothermal source of heat, close underground and maybe volcanic. The most famous of them, Old Faithful at Yellowstone National Park, erupts about every hour. The theory of the geyser was expounded by Robert Wilhelm Bunsen in the nineteenth century, and I liked the idea of making an artificial one for scientific television. My RIG mused on the problem of how to make one out of domestic materials that would appeal to a TV audience. I chose a metal electric kettle as the heater, joined to an elevated trash can lid by a glass tube.

I planned a tube about 2 meters long and 4 centimeters in internal diameter. That 2 meters of length made practical sense—it was about the longest glass tube I could handle or transport. The 4 centimeter diameter was a miracle of physical intuition. I had no idea how or if the geyser would work and just guessed an effective tube diameter. For strength and rapid assembly, I planned to put the whole thing in a stout frame of steel Speedframe tubing. Speedframe is a hollow steel section, about 2.5 centimeters on a side. It can be cut into lengths, and its corner pieces allow those lengths to be assembled neatly and strongly with plastic inserts. The whole structure can be knocked apart with a mallet and built again.

Water in the vertical pipe would pressurize the boiler below. Its water would therefore boil above normal temperature. The resulting steam would occupy a huge volume and blow the water out of the vertical tube. The sudden loss of hydrostatic pressure would leave the remaining water in the kettle superheated. It would boil vigorously and blow further steam and water up the tube and out of the geyser, enhancing the upward jet. That jet would cool and fall back into the trash can lid, condensing the steam and filling the whole thing up again. The kettle, still on, would soon boil the returned water for the next eruption. My RIG imagined that it would imitate a real geyser.

So I began to build. I soldered the kettle lid on tight and made a hole in it for my glass tube. I soldered a short copper central-heating pipe "collar" onto it, to fit that tube—made for me by the Newcastle University chemistry department glassblowers. I sealed the spout of the kettle with a turned-off tap. At the top of the tube, I put the metal trash can lid, again with a hole, again with a soldered copper collar. Very cleverly (as I thought), I fastened all the plumbing together with bits of rubber bicycle inner tube of about the right size, held on with wire.

The whole monstrosity was a triumph of guesswork (fig. 6.1). I summoned my courage, filled it with water and switched it on, but did not know what to expect. Even if the thing showed geyser action, how long would it take to start and how long would there be between eruptions? If it gave trouble, I might be able to make a few modifications, but not many. I could perhaps have increased the density of the fluid, by filling my geyser with salt water, and I could easily reduce the electrical power to the kettle. I could also control the depth of the glass tube in the kettle; the prototype had it as large as I could make, some 10 cm. Amazingly,

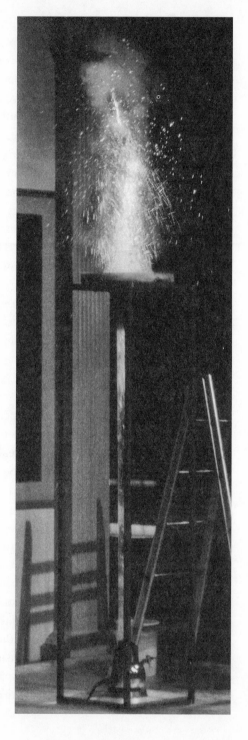

FIGURE 6.1
Artificial Geyser at Work

The basic heating unit is a metal electric kettle. Its spout is sealed by a turned-off tap and its lid is soldered in place. That lid is connected via copper pipe and bicycle inner tube to a glass tube, with a length of 2 meters and a width of 4 centimeters. The top of the tube is connected, again by a soldered copper tube and a bicycle inner-tube fastening, to a metal trash can lid. The whole thing is filled with water. When the geyser erupts (about every minute), much of that water is upflung about a meter out of the top of the glass tube. Most of it falls back into to the trash can lid and drains back into the electric kettle. The artificial geyser works best when the glass tube goes well into the kettle. The ladder on the right allows the geyser operator to fill it with water.

the whole thing worked on test. It took about 5 minutes to warm up, and then every minute it ejected about 2 liters of water upward, in a dramatic jet.

In the YTV studio, about 80% of the water fell back into the lid and drained back into the geyser. The rest spread onto the floor of the studio, where it annoyed the cleaners. After about five eruptions, the monster had lost so much water that it needed refilling. Those clever bicycle inner tube joins were leaking, too. Luckily, I had brought down some tightening wire and could just about resecure them.

The whole lash-up lasted long enough to show a studio audience and to record material for the show. But unlike Old Faithful, it was highly unfaithful and did not match my RIG's dream of a display that you could just leave going.

The Steam Balloon

Steam, or water vapor, is much lighter than air. It's hot, too; it shares the buoyant advantages of hot air. So in the 1960s, as a young chemist of about 24, I tried to make a steam balloon. Any balloon contains some gas less dense than air. There are few potential balloon gases. The favorites are hydrogen and helium; then we have methane, ammonia, steam, hydrogen fluoride (possibly), and neon. Hot air is a specialized medium: in practice, it is continuously regenerated as fast as it cools and escapes. Hot steam is specialized too. It condenses to water. My early design carried a reboiler to vaporize that water again.

The general rule for making balloons is this: the bigger the better. Double the size of the design, and its weight goes up four times. But the volume of gas it will take goes up eight times. You get double the lift for the weight. By contrast, heavier-than-air flying-machines get harder to make as they get bigger. Anybody can make a paper airplane; but even a plane to lift one man demands serious technical skill!

Anyway, I felt enthusiastic about my steam balloon. Its envelope (the balloon part of the contraption) had to withstand hot steam and to be a good thermal insulator. My envelope was two sheets of polypropylene, a plastic sheet that melts way above the temperature of steam, about 0.025 millimeters thick. Between the two films there was to be a thermal insulator, composed of tissue paper and lots of air. I devised a gadget for sealing the envelope film into whatever shape I chose. It made a sort of three-layer "sandwich" (fig. 6.2). My intuition felt that it would be adequately light and would retain heat well.

I made that balloon as big as I could handle—about 1.5 meters high and 2 meters in circumference. I gave it a supporting frame of stiff wire for the top and a conical bottom to hold the reboiler. I hung that wire frame in my parent's garage where I was building the balloon, but I hoped that when I filled it with steam it would lift from the frame.

Then I tried it out! Into that conical bottom I passed steam from a tin boiler, driven by a Primus stove burning kerosene. Even with the stove going full blast, I never managed to fill that balloon with steam, and it certainly never lifted. The steam inside just condensed to water, which drained back into the boiler. I had totally underestimated its rate of condensation. Had my intuition been wrong?

After that defeat, I left the idea alone for decades. But in my TV activities, I came across a new plastic film, aluminized polyester (Mylar), and revived the idea. Pocket bags of it are sold to rough-country walkers. The film is only 0.01 millimeters thick, so quite a big bag can fold up small. Furthermore, the coating is very shiny. If a hill walker gets lost and benighted, he can take a bag out of his pocket, open it up and get inside it. He conserves body heat, and his location is obvious to a search party.

I liked the idea of making a steam balloon from this new thin film. Aluminum reacts with steam, so all the aluminized surfaces of my planned balloon would have to be outside. And making it would be difficult. Polyester film may resist steam, but it is very hard to join. The 3M company told me of a double-sided sticky tape that could join it and could even withstand live steam for some time.

So, after a long interval, I started balloon tailoring anew. My new balloon was much the same size and shape as my old one: about 1.5 meters high and 1.5 meters around. It was made from my new polyester film, aluminized and 0.01 millimeters thick. Again, in my fantasy I hoped that this balloon would lift itself and a reboiler. I even built that reboiler, from thin-wall aluminum containers.

I tried the initial design in my kitchen. I blasted steam into it from a 3-kilowatt electric kettle. It did not even lift itself, let alone the reboiler. I had to admit that my fantasy would never work. Again, I fear that my intuition was at fault. So I simplified my scheme. I abandoned the reboiler.

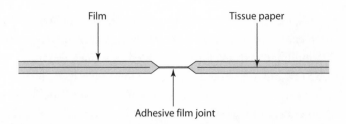

Film   Tissue paper

Adhesive film joint

FIGURE 6.2 Cross-Section through Fabric of Prototype Steam Balloon
Polypropylene plastic film about 0.025 millimeters thick, joined at intervals by
hot-melt adhesive coating. Tissue paper sheets separate the films. This construction was intended to limit heat loss from the balloon.

An aluminized polyester envelope filled with steam should rise, at least
until the steam inside it condensed. My RIG began to imagine a clever
"steam trap" to let out condensing water while retaining steam.

I even stopped relying on my intuition. I set up an experiment to find
how fast heat was lost from a steam-heated surface. Even my aluminized
polyester film lost about 300 watts per square meter. No wonder my
steam-balloon experiments had failed! Now with reliable data, I could
calculate the best size for my polyester envelope. It was about 60 liters,
maybe a half meter across. I didn't bother with my clever filter valve. Even
if I invented one, it might be too heavy. Lightness was the key!

The tailoring of that balloon posed geometrical problems. My sticky
tape was both heavy and weak. I wanted to minimize the length of the
seams, while creating a fairly spherical balloon from flat film. My final
design was basically pentagonal. But when inflated with steam from an
adapted electric kettle, it made a fine balloon. I could switch on the kettle,
and, when I judged that the balloon was about full of steam, I could push
it off the filling nozzle. (I wore gloves, letting me handle hot objects without fear of being scalded.) Once the balloon was off the kettle, a neat little
elastic-band gadget on it snapped it shut to retain the steam. On TV, in
front of an audience, everything worked. My steam balloon stayed up for
25 seconds, or even more.

In my trials with this technology, I noted (as had James Watt, see
chapter 1) that steam condenses much more slowly if there is a little air
in it. Fortunately, there was always a bit of air in my balloon.

Later still, I greatly improved my steam balloon technology. I started to play with cheap kitchen trash bags. They are white polythene film and only 0.005 millimeters thick. Each is a sealed container about 75 centimeters long, 55 centimeters wide, and 8 grams in weight. It has a seam at one end, and you can seal the other end with a single strip of sticky tape. That tape will hold steam for a minute, and you can even add an attachment valve. I have not even bothered to measure the melting point of a trash bag, nor its rate of heat loss or its area or volume. But I can fill one with steam from my kettle, and it rises nobly as a balloon!

That demonstration enhances any lecture. It takes an ordinary domestic object and does something quite unexpected with it. Then you can take the argument in any number of directions, from meteorology to air travel. In this latter case, the steam balloon invites comparison with the hot-air balloon. But quite apart from its technical problems, steam is far inferior. The modern hot-air balloon depends on polyester fabrics for its envelope and liquefied propane or butane for its fuel. By contrast, my long struggle with steam as a lifter showed me (first) that calculations and tests have a mighty edge over intuitions and (second) that hot air is a much better thermal lifter anyway. Even so, I do not regret my lifelong entanglement with the idea!

7

Astronomical Musings

*P*ulsars are tiny dense stars about 15 kilometers across. They are not quite the black holes that Subramanyan Chandrasekhar wondered about (see chapter 1), for they emit a radio signal. They rotate about once a second, and their signal repeats at that frequency. Such a signal was first observed in 1967 by Jocelyn Bell Burnell, a student of Tony Hewish's. She helped to build the Mullard radio telescope that saw the signal, in Cambridge, United Kingdom. Was it real? The telescope was new and readily picked up terrestrial interference. As with any new observation or RIG idea, the scientists did not wish to make public fools of themselves. It took months, and many checks, before they summoned the courage to announce their discovery. Few scientists want to risk looking like a fool; yet it is part of having ideas or chasing up observations. Here are some notions of mine, also on astronomy, also possibly foolish.

Parallel Universes

We can see only one universe—our own. And it seems governed, not by law, but by quantum probability. Why anything actually happens, you are not supposed to ask. Einstein did not like this at all. "God does not play dice," he famously said. But modern physicists put up with it. Uncertainty even worried Isaac Newton. He knew that when light went through glass, most is transmitted but a small amount is reflected (which is why you can often see a weak reflection in a glass window). Newton saw this as a matter of timing. About 96% of the light was transmitted during a burst of transmission, while about 4% was reflected during a shorter burst of reflection. These days we just blame quantum probability. An incident photon has a 4% chance of being reflected and a 96% chance of being transmitted.

Such probabilities were brutally exorcised in the 1950s by Hugh Everett III. He proposed that every time a quantum uncertainty came up, the universe split into two: one in which it happened and one in which it did not. The universe we inhabit is the result of all the quantum events of the past. They now have their historical values: they either happened or they did not. But what about all the other universes, which made different quantum choices? Ah! said Everett, they are still there, but we can't detect them. And each time a quantum choice comes up, which it does many thousands of times a second, our universe splits into two again, and we split along with it. In one such Everett universe, for example, radioactivity has never occurred. Radium and plutonium are ordinary geological minerals and items of commerce. In another, all radioactivity occurred long ago, and only stable isotopes, such as iron-56 and sodium-23, now exist. In yet another, light goes through glass without reflection. The bloomed lens to reduce surface reflection has never been invented: there is no need.

Everett's bold theory is not well regarded. J. S. Bell (the physicist who originated Bell's inequality) said of it, "If you take it seriously, it is hard to take anything else seriously."

In 1998 Daedalus proposed that the multiplicity of Everett worlds must "add up."[1] A radioactive decay in one world implies a continuing atom in another. A successful lucky chance in one world must fail in another. The notion fitted my suspicion that the quantum events we observe are not quite random (so that, for example, radioactive half-lives are a bit bogus; see chapter 16). Another bit of evidence is the Schmidt machine.[2] In this device, four electric light bulbs are lit in a allegedly random sequence by the decay of a radioactive source, strontium-90. Mr. Schmidt claims that certain observers can guess, with better than chance probability, which bulb will light next. Some can even *will* a specific bulb to light next, with better than chance probability. This device has not yet been replicated or written about in the refereed scientific literature, but I like the idea. Again, it suggests that our hallowed quantum choices are not quite random. Some human decisions, spanning the universes by some sort of telepathic power, can outguess them.

Daedalus even had evidence for this claim. He recalled a study of intuition in business executives.[3] Successful ones scored better than chance; but failing ones scored worse than chance. Clearly these perverse "antipsychics" were, sadly for them, tuned to the wrong world. Daedalus hoped to assemble a panel of people who can guess wrong. He planned to find such rare and gifted individuals from among bankrupt businessmen, failed spiritualists, and inspired losers of all kinds. He intended to look for statistical agreements among their hopeless fantasies.

Anyway, said Daedalus, these bits of evidence imply another world to make the numbers add up. Somebody, somewhere, may be signaling to us! They are making our quantum events nonrandom! A vast field of research opens up: the study of phenomena that should on quantum principles be entirely random, but in practice are not, or not quite. Radioactive decay is an obvious example; electrical circuit noise is another; photon counting of optical beam-splitters is another. Applied to the results, modern methods of decoding should reveal the messages that the physicists of other Everett worlds are transmitting to us. Any positive outcome would be revolutionary.

And where are all these invisible Everett universes? In principle, an extra dimension could make room for any number of them (and string theorists have posited 10 or 11 dimensions compared with the three of our own universe. I discuss extra dimensions later in the chapter). Daedalus has imagined those invisible universes in such an extra dimension. A parallel universe could be less than a millimeter away from ours, yet be undetectable and untouchable in another dimension. Just as well, too, since half of those universes might be anti-matter, from an early "big bang" quantum choice. Normal matter has positive nuclei, with negative electrons going around them. Anti-matter is the reverse. It would react violently with normal matter, and none is known experimentally. But it might exist in another dimension. That extra dimension gives a whole new volume, just as a book can have many closely packed pages, with separate information on each page. I fantasized that "angular" universes might couple between the parallel, physically real, but non-communicating Everett worlds that have made well-defined quantum choices. That angular coupling might be a "ley line" (fig. 7.1). Some believe that ley lines—invisible lines of psychic power and energy—exist on the Earth. I

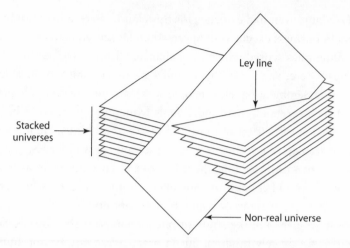

FIGURE 7.1 Universes Stacked in a Space of One Extra Dimension
The flat sheets represent two-dimensional universes stacked together in a space of one more dimension. They are bisected by a non-real universe. This is permitted in some theories, but the resulting "ley line" is advocated only by mystics.

know of no attempt to measure a physical constant (such as the speed of light) along one.

Even I can see no way in which a parallel universe could exchange material with us, as well as information. Daedalus, however, has wondered whether its inhabitants find extra socks in the wash but suffer mysterious losses of wire coat hangers.

Daedalus has also mused that parallel universes contain intelligent life and indeed, that they have physicists who are signaling to us. So far, we have not detected other intelligent beings even in our own universe. The Arecibo radio-telescope with its Project SETI (Search for Extraterrestrial Intelligence) has invited users of personal computers to leave their machines on and have them programmed to scan a narrow band of the recorded Arecibo signal. SETI is looking for some sort of nonrandom component in that signal. Many thinkers have imagined the likely form of coded messages from other civilizations—lists of prime numbers, the digits of π, and so on.

If I were an alien civilization out to spread the Word, I would not bother with radio. I'd modulate the local star. (Don't ask me how.) As far as I know, nobody has looked for high-frequency modulations on the visible light put out by any star, except for a few amateurs looking for eclipses of a Saturn-like planet with rings. As the rings go in front of the star, its light should show brief feeble rapid fluctuations.[4] The chance of seeing such an event seems very small. A planet like Saturn only has rings for a few million years. Ultimately the small moonlets that make up the rings must fall down onto the planetary surface.

I share J. S. Bell's unease with Everett's theory. Why postulate vast numbers of unobservable universes, just to allay the puzzles of quantum mechanics? Occam would be outraged! Even now, quantum mechanics continues to make correct predictions but defies philosophical understanding.

Lots of Dimensions

Peter Stubbs of *New Scientist* told me of this exchange between two scientists at a Liverpool scientific conference in the 1970s:

RONALD GIRDLER (Newcastle upon Tyne): We who are close to Sir Edward Bullard believe that . . . XXX.

JOHN CLEGG (London): I assume the effects of Sir Edward Bullard fall off as the square of the distance? We at Imperial College believe that . . . YYY.

Light, sound, electric and magnetic forces, Newtonian gravity, and (possibly) the effects of Sir Edward Bullard, all decline as the square of the distance. This implies a universe three-dimensional on the scale of the measurements. Many natural forces seem to fit, and engineers agree. Accurate drawings show technical objects in three dimensions: front to back, side to side, and top to bottom. And asymmetrical technical objects such as screws and gloves exist in right-hand and left-hand forms. They can't be interchanged. A four-dimensional world might have its own sort of handedness, but three-dimensional screws and gloves could simply be flipped over in it.

Some cosmologists, however, already imagine an expanding four-dimensional universe. A common analogy has two-dimensional galaxies drawn on a rubber balloon. Blow the balloon up, and all the galaxies move away from each other. If you add a dimension, the galaxies become three-dimensional. The universe, as the expanding balloon, becomes four-dimensional. Einstein's notion that space may be slightly bent makes more sense with a four-dimensional space to contain that bend. With time as yet another dimension, we may have five in all. Our universe may even have a Möbius twist to it, suggesting yet another dimension. If so, an astronaut who went all the way around our universe would come back totally inverted. The astronaut would have a heart on the right side of the chest and would depend on packed provisions now containing right-handed amino acids. The left-handed molecules of normal food would now be indigestible. Daedalus's interpretation of Everett's notions implies four dimensions. He points out how a whole new vast collection of Everett universes could neatly be fitted into one extra spatial dimension. So how many dimensions does space have?

Let us imagine that on a large-enough scale it is four-dimensional. Nobody seems to have thought about the transition and whether it starts detectably in the solar system. The Pioneer space probes—which were sent into space in the early 1970s and observed Jupiter before continuing to travel out of the solar system—are free-falling objects. They are going rather more slowly than current gravitational theory would expect. Conversely, spacecraft sent around Jupiter may gain a little more speed than calculated. Are these deviations related to the dimension of the space they are exploring?[5]

Dimensionality seems to matter on the small scale too. String theorists want 10 or 11 dimensions of space to hold their strings and allege that all but three of those dimensions are curled up so tightly that we cannot detect them on any human scale. Strings exist (if they do) on the scale of the Planck length, 10^{-35} meters, much smaller that the 10^{-15} meters of the electron or the 10^{-10} meters of chemical molecules.

Incidentally, string theory itself may have started in a telephone call between the great physicists John Wheeler and Richard Feynman:

WHEELER: Why do all electrons have the same charge?
FEYNMAN: I don't know. Tell me!
WHEELER: Because they are all the same electron!

Wheeler imagined a snaky electron in four dimensions repeatedly passing through our three dimensions. Wherever it appeared in our space, it did so as a sphere. Wheeler was imagining a four-dimensional world,

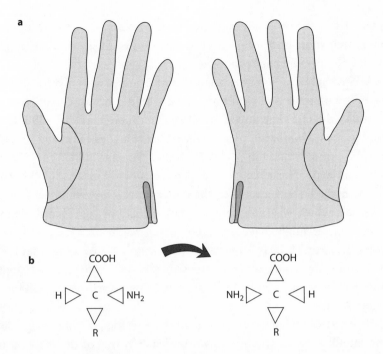

FIGURE 7.2 Gloves Are Like Molecules
Gloves (*a*) can be left-handed or right-handed. They cannot be inter-converted in our three-dimensional space. Some chemical substances (*b*) have molecules that also cannot be inter-converted. (The triangle denotes an ordinary chemical bond with the wider end nearer the eye.) The amino acids found in living things are almost all left-handed. In the pictured example, R = CH_3 gives the amino acid of life, alanine. Molecules with "handedness" can sometimes be slowly inverted by a process called "racemization," indicated by the pictured arrow. Nobody has devised a way of doing it with gloves.

which string theorists have taken to extremes. Yet on the small atomic scale, even chemists prefer three-dimensional molecular models. They even allow that some molecules, such as the amino acids of life, have handedness, like gloves. Over time, however, they can slowly lose that handedness, a process known as "racemization." Given enough time, half of the molecules become right-handed (fig. 7.2) Indeed, racemization is used to date old biological specimens. It has been proposed as a way of dating the Dead Sea Scrolls[6] and has been used to study the age of bear dentin.[7] I fantasize that the molecule briefly accesses the fourth dimension. In that short time it flips over and returns to the three-dimensional world with opposite handedness.

Accordingly, you might expect the dimension of the universe, and its variation with scale and with time, to be a matter of hot debate. Not at all: in fact the last time anyone tried to determine it was in 1790. Karl Gauss proposed, and even tried, to measure the angles of the big triangle formed by three mountain peaks. He found that they added to 180°, as you would expect for a flat three-dimensional world in which light travels in straight lines. Gauss used the conventional surveying technology of the day, which was probably accurate to 0.1° or so.[8] Had he found a deviation, he might have upset physicists, mathematicians, and even theologians. For Thomas Aquinas once drew up a list of things that even God could not do.[9] One of those impossible things was to make a triangle whose angles do not sum to 180°. We can do the experiment much more accurately now.

So my RIG imagines a new big experiment (fig. 7.3). Let us put up three satellites in geosynchronous orbit, each at one of the three vertices of an equilateral triangle, and let us reflect a laser beam around the triangle. It could all be done with mighty modern precision. If the laser beam goes all the way around and closes accurately on itself, the three-dimensionality of earthly space would be triumphantly established. If it is even slightly wrong—and current interferometric methods could define the error with great precision—physicists and mathematicians would have a whole new field to argue about. Geosynchronous satellites, each some 42,000 kilometers from the center of the Earth, would give a triangle with a total optical path of 218,240 kilometers. A geosynchronous orbit seems familiar and convenient, but some other orbit might be technically preferable. One satellite will carry a laser to launch the beam and a detector to measure it coming back. Each satellite will have an adjustable wedge-shaped plate to measure and correct small deviations from the expected 60° of bending. It will take light about 0.73 seconds to get around the triangle. So for an interferometric study, the laser should emit continuously. A pulsed laser would be quiescent by the time its pulse returned.

Yvan Bozzonetti of Paris (a correspondent of Daedalus) has quite another way of judging dimensionality. He points out that ordinary gas

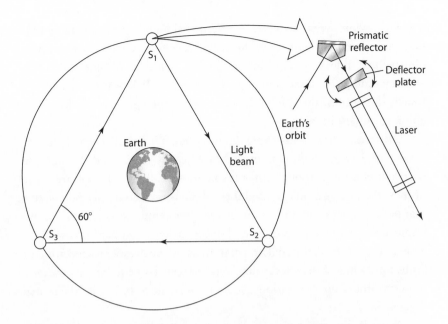

FIGURE 7.3 Proposed Satellite Experiment to Check Space Flatness
Three evenly spaced satellites (S_1, S_2, and S_3) are in orbit around the Earth. A
light beam from a laser on S_1 is reflected around the satellites. The light beam
should form an equilateral triangle with angles of 60°. Each satellite carries
prismatic reflector a bearing a metallic reflector and shown in more detail for S_1.
If the space explored by the three satellites is not "flat," a deflector plate on each
satellite can alter the angle needed for the light beam to go all the way around
the triangle.

molecules have a thermal energy of $(1/2)kT$ for each degree of transla-
tional freedom, where k is Boltzmann's constant and T is the absolute
temperature. In three-dimensional space, therefore, they have thermal energy of $(3/2)kT$. If space expanded to four dimensions, they would each
need thermal energy of $(4/2)kT$, and so would lose $(1/2)kT$ of energy. If
their initial temperature was 300°K, about room temperature, they would
cool to 225°K or −48°C. A change in temperature can easily be measured
to within 0.001°C. So any experiment to vary spatial dimensions (such
as a Pioneer-type space probe) should look at the temperature of the gas
molecules in it. A change from 3.0 to 3.00004 dimensions could readily
be spotted.

Meanwhile, what would a four-dimensional world look like? Our first study should be nature. Evolution never seems to use more than three dimensions: unless the replication of DNA, without all that unwinding and rewinding, is an example (see chapter 16). If DNA can access the fourth dimension, the double helix could simply peel apart, duplicate, and reform within it.

Otherwise, life seems three-dimensional. Birds still keep their embryos safe in eggs, although an egg could easily be entered through the fourth dimension without even cracking it. And all living things are made of cells, each being a small closed three-dimensional container. So it seems that nature ignores the fourth dimension. Yet evolution works by the extreme modification and extension of one idea, leaving other ideas severely alone. Thus life has created flying mammals by the drastic modification of limbs into wings. It has never invented the wheel, or fire, or microwaves, or many other human creations. So it may have done little with the fourth dimension.

Accordingly, my RIG is musing on the revolution that four-dimensional technology could bring. Let us assume that on the tiny scale of the Planck length, 10 or 11 dimensions are easily accessible. On the larger molecular scale, a volume the size of a molecule might briefly access the fourth dimension every few thousand years or so—agreeing with the slow racemization of old specimens. With a lot of amplification, a useful volume of four-dimensional space could perhaps be accessed long enough for a useful technical operation to be carried out in it. The implications would be dramatic.

The first revolution of the fourth dimension would be medical. No longer would the surgeon have to make an incision and push intervening material out of the way. In four dimensions, the field of operations would be laid out completely, like a two-dimensional map in three-dimensional space. The surgeon could get at any point of the body directly, through the fourth dimension, do the work required and not even leave a scar. A woman giving birth would not need to balance bodily damage against the safe emergence of her child. She could give birth through the fourth dimension. Her baby would be delivered with no stress to itself and no stretching or damage to her. In a sense, caesarian delivery would become universal. It would cause no pain or trouble.

Fabric too would be revolutionized. Only in three dimensions is a knot possible. So string as a confiner and rope as an anchoring material would not be safe. Any knitted garment could be made or undone through the fourth dimension. Looms and knitting machines, like most other engineering masterpieces, would become obsolete. Indeed, the most intricate assemblies would become absurdly easy to engineer. The question often posed by amateur repairers, "How the hell did they ever put this thing together?" would be even harder to answer. Can openers would become obsolete; you would enter the can through the fourth dimension.

Crime and punishment would be transformed as well. No safe or strongbox could keep valuables secure; any thief could get at the contents through the fourth dimension. And even if apprehended, a thief could escape from jail. The criminal would simply walk out through that dimension. Worse still, armor would become useless. No tank could protect a soldier against an attacker; any malefactor could put a bit of metal or poison in any part of the body without even leaving a mark. A whole new sort of four-dimensional security would have to be invented.

Screw and glove businesses would be much simplified, too. You would only need to make one form and turn half of them over in the fourth dimension, as is perhaps implied by paleochemistry. I am reminded of the trick invented by Sam Goldfish, who around 1900 became the top salesman for the U.S. Elite Glove company. He bought gloves in France and split each pair into its left-hand and right-hand member. He bundled the sets together and sent the left-hand set to one address in the United States and the right-hand set to another. He failed to pay for the shipping charges. The carrier therefore held onto the two glove consignments and later auctioned them off. Nobody wanted a set of single-hand gloves, so Goldfish was the only bidder at each auction. He bought both sets and matched them up in pairs again for resale. Later he changed his name to Sam Goldwyn and became a leading light in the film business, in fact the G in MGM. Four-dimensional technology would have frustrated the trick that got him started.

Perhaps the most romantic description of four dimensions occurs in the novel *Flatland* by Edwin A. Abbott.[10] All the characters in the novel are geometrical figures and much of the action takes place in a two-dimensional world called "Flatland." Abbott imagines an unfortunate square, imprisoned by the ruling aristocratic circles for asserting that there are three dimensions. The square is thoroughly out-argued by the circles; yet he has visited Spaceland and has seen the third dimension for himself. The creatures of Spaceland look down on his flatness but deny with horror the idea of a fourth dimension.

Mainly about Carbon and Hydrogen

Edward Wheeler once left some packaged bacon in my briefcase in error. He then imagined me finding it and inventing a theory of the continuous creation of packaged bacon. This notion is not totally absurd. The theory of "continuous creation of hydrogen" was espoused and developed by Fred Hoyle as an explanation of the universe. In this theory, hydrogen appeared all the time throughout space. The process neatly counterbalanced, and indeed drove, the expansion of the universe. In this theory, the universe had no beginning and has always looked much as it does now. Cosmologists seem ultimately to have rejected the continuous creation of hydrogen (to Fred's fury). Thanks largely to the cosmic microwave background, they have adopted an alternative big bang theory—Fred's term—in which the whole universe started about 20 billion years ago. Daedalus has reacted to the theory of continuous creation of hydrogen, once by proposing the Albert Hall in London as an ideal vacuum desiccator for testing the appearance of hydrogen in it and later by querying the theory.[11] He asked, what velocity does that hydrogen have when it appears? If you found an inertial frame in which the hydrogen was created stationary, it would in effect be a privileged, static frame, contrary to relativity.

This leaves the chemical composition of the universe unexplained. Suppose continuous creation made hydrogen, or the big bang made energy that soon condensed to hydrogen. Where did the heavier elements come from? Stars get and stay hot because of a fundamental nuclear fusion reaction, four atoms of hydrogen going to one of helium. William Alfred Fowler, Margaret and Geoffrey Burbidge, and Fred Hoyle took the notion further.[12] They proposed that all the heavy chemical elements were formed inside stars by nuclear reactions. These elements were later blown out into the interstellar gas when the stars exploded (as many do: they become novae or supernovae). They squared this theory with the elemental composition of the universe, as far as it is known. So the interstellar gas has heavy elements in it. These are obvious in the planets we know (geologists reckon that the Earth has an iron core, for example).

James Jeans's theory of planets made them very rare—they could only form if two stars approached each other closely. Later theories made them much more common. As a star condensed from the interstellar gas; planets formed along with it. Yet the very first stars, condensed from new big bang interstellar gas, might not have had planets. Or they might have been gaseous hydrogen-rich ones, quite unlike the planets we know. Only second-generation stars, formed from interstellar gas already enriched with heavy elements from exploding first-generation stars, could form solid planets like the ones in our system. Incidentally, if a star collapsed into a black hole, as in Chandrasekhar's musings (see chapter 1), its mass would not change. Any planets it had would continue to orbit it. Planets seeming to orbit nothing have not yet been detected—indeed without the light of a central star they would be very hard to detect—but they would be powerful evidence for Chandrasekhar's notions.

Explosions are not the only way in which heavy elements might get out of stars and into the interstellar medium. Even stars that do not explode may push out heavy elements continuously. Our own sun emits a solar wind mainly of protons and electrons, that is, ionized hydrogen. Some red giant "carbon stars" emit a wind that contains a lot of carbon. Like all stars, carbon stars are very hot and must be even hotter inside. They seem to emit solid carbon and some of its compounds. Carbon forms a wide variety of solids, which may possibly include "amorphous carbon" (see below). Oddly, no liquid seems to have ever been reported anywhere.

I do not know if any of those carbon solids are stellar products. I love the idea of a carbon star surrounded by diamond planets, each growing steadily by the capture of diamond from the star. (Daedalus once invented diamond fiber and advocated it for making lingerie for lady spies—lustrous, glamorous, bulletproof.[13]) But even a carbon star probably emits a "stellar wind" that is not pure carbon. There will be a lot of ionized hydrogen in it. I imagine that the molecules that cool from that wind will be hydrocarbons: methane, ethane, benzene, and so on. Any planets that result will not be diamonds but oily blobs. This argument supports

Tommy Gold's notion that geological hydrocarbons are ancient natural products. (An alternative theory is that, like coal, all fossil hydrocarbons are the remains of once-living matter.)

My interest in carbon arises partly because Daedalus proposed the hollow carbon molecule 20 years before Robert Kroto, Harold Curl, and Richard Smalley won the Nobel Prize for chemistry by making buckminsterfullerene.[14] Daedalus has also mused on those strange explosives, cuprous acetylide, silver acetylide, and gold acetylide, which consist entirely of metal and carbon. So what do they explode into? Daedalus has suggested that they explode into diamond. He has proposed silver acetylide and gold acetylide as a way of making instant silver-diamond or gold-diamond jewelery.[15] But the sudden explosive creation of carbon is more likely to give "amorphous carbon," a form of the element that seems to have no microstructure. Later I mused that this may be a randomer (see chapter 13).

As a chemical, carbon seems rather unexciting. It is a black solid, most easily available as charcoal. Charcoal filtering to remove colored impurities is a known chemical technique, which I have shown in lectures and TV demonstrations. For such schemes, I got activated charcoal from the Northern Carbon Research Laboratories in the Newcastle chemistry department. They gave it to me as a coarse powder. Activated charcoal is a powerful absorbent and is made by treating ordinary charcoal with steam at 1000°C. The species of wood matters: beechwood is good. Activated charcoal is used, for example, in gas masks, for absorbing many noxious materials. I have also tried making my own activated charcoal, by burning bread in an oven at 1000°C. The result was useful on TV where the producer wanted something familiar. We called it "burnt toast." I helped it along by making it from "sandwiches" of activated charcoal. Yorkshire Television Ltd. used my burnt toast to show how charcoal decolorizes tea, brown sugar, red wine, and so on. To "grab" the TV audience, we decolorized and ruined a very expensive red wine.

There is an important lesson here: to seize an audience, you need the equipment to be familiar; only the outcome should be dramatic. I once showed a charcoal decolorization in a public lecture. I turned red wine the color of clear water. The audience was highly perturbed by my drinking the decolorized wine from a chemical filter flask. So in my next lecture I used domestic glassware. The audience accepted it easily and paid attention to the chemistry. This lesson stayed with me and greatly influenced my TV work. Use domestic equipment if you can!

How Far Away Is That Star?

This is the basic problem of astronomy. There are many ways of finding out, but the astronomical "gold standard" is the determination of stellar parallax, the slight yearly shift in the apparent position of the star as the Earth orbits the sun. Daedalus once found himself thinking about it.

The stars are very far away, and stellar parallax is extremely small. Indeed, in the Middle Ages, its absence was a powerful argument against the whole Copernican theory that the Earth goes around the sun. Modern astronomers consider that the stars are suns themselves but very far away. This theory was first put forward in the Middle Ages but offended theologians. Parallax was finally found in the nineteenth century by very precise telescopic measurements on relatively near stars. It went on to become the most exact way of determining the distance of the nearer stars. Indeed, the Hipparchus satellite was put up in 2000 to determine a precise stellar parallax for each of a large number of stars. Once parallax has been determined for a given star, the distance of other stars of the same type can be determined, perhaps by comparing their brightness with that of the known standard star.

Daedalus once put forward a new way of determining stellar distance. It was based on brightness. Like parallax, it depended on the fact that the Earth orbits the sun every year. Daedalus reasoned that for half of the time the Earth was nearer the star, by about a radius of the earthly orbit; for the rest of the time it was farther away. The star should be brighter when it was nearer the Earth and fainter when it was more distant. So it should show a yearly fluctuation in brightness (fig. 7.4)!

Now brightness can be determined absolutely, by photon counting. I was very familiar with photon counting. Many Raman spectrometers use it. As a chemist I did a lot of Raman spectroscopy, and I felt good about photon counting. A telescope detects photons from a distant star. If I could calculate how many, I could get some idea of whether my idea was feasible.

Like most astronomers, I took a star to be a black body, which absorbs every photon of light that hits it. The way a black body emits pho-

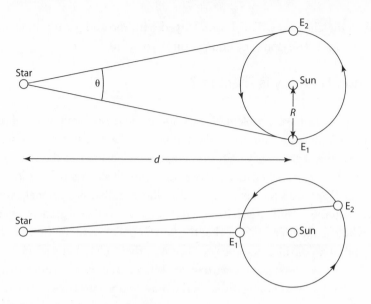

FIGURE 7.4   How Far Away Is That Star?
The right-hand circles show the Earth orbiting the sun in a circular orbit of radius R. It does this every year. The standard way of measuring a stellar distance (d), is to take two photographs about six months apart (E_1 and E_2) and determine the angle Θ from them. This is the "stellar parallax" method, shown in the upper part of the picture. Daedalus's scheme is shown in the lower part. He notes that the star is brighter when the Earth is nearer to it (E_1). Six months later it is farther away (E_2), and the star should look fainter. So he feels brightness measurements, "photon counting" should work as well.

tons is well known, but I could not manage the calculation. So I wrote to David Whiffen, a powerful physical chemist and once head of chemistry at Newcastle University. He had recently retired but still had a ferocious understanding of physical principles.

In my previous discussions with Whiffen, I had felt myself utterly outclassed. He would listen to my concerns and then start talking about something entirely different. Oh dear, I would think, I have not stated things clearly enough. Not at all. From some distant site of physical theory, Whiffen was building an intellectual structure which in half an hour's time would come down on my silly problem and solve it and all related problems in full generality! That was how he thought. My main concern was not to look too stupid in front of this terrifyingly acute intellect. I developed a defensive strategy. I would say at random, "Shouldn't there be a 2π in that somewhere?"—giving the false impression that I was following his argument. David would then halt and consider whether there should be a 2π in his expression or perhaps whether there was one already. Meanwhile, I had a chance to puzzle out what the whole expression was about.

Whiffen soon solved my stellar problem. His basic result was that for every square meter of its surface, a black body radiates $15.1(T^3k^3/h^3c^2)$ photons a second. Everything is in scientific MKS units: the Meter-Kilogram-Second units in which all science is done. T is the absolute temperature of the surface, k is Boltzmann's constant, h is Planck's constant, and c is the speed of light. Using this formula, the sun puts out 1.78×10^{45} photons a second. An average star (a bit brighter than the sun) might put out 10^{46} photons a second. I had reached a fairly similar conclusion myself, having made a wild approximation that turned out much better than I had any right to expect.

This implies that a big telescope receives perhaps 6 billion photons per second from a star close enough for its distance to be measureable. Of these, perhaps a hundred million are optical or at least can trigger a photomultiplier and work a photon counter (these instruments measure not the mere brightness of light but rather its individual particles—photons). The problem is complicated by the fact that starlight is not steady. It is highly "noisy." To make a good distance measurement, you need to count so many photons that the noise averages away. My calculations suggested that for a star 100 light-years away, you needed to collect and count about 10 million million photons, which means counting for rather more than a day. If you count for all the available time, say, for 6 months, you can measure distances out to about 1,200 light-years. This is a bit farther than the reach of conventional stellar parallax photography.

The great advantage of brightness measurements is that you can get away with a bad telescope. In all probability there will not be another star near the one whose photons you are counting. So it doesn't matter if its image is a bit blurred or unfocussed. You just count all the photons coming from that patch of sky. The Daedalus column assumed a rather imperfect space telescope. Later, we got one without even trying!

The advantage of stellar parallax is that it is very quick. Two photographs, taken 6 months apart, should do the trick. But it pleases me that Daedalus, the non-serious theorist, was able to come up with a new method of determining stellar distance and even a new astronomical technique.[16]

Later I mused that the astronomical telescope has been developed over several centuries, essentially for one purpose only. It has to determine the position of an object in the sky and to resolve it from other objects nearby. Furthermore, each advance in telescope design has brought about astronomical advances, too. I am thinking here of new optical regions opened by new astronomical methods—ultraviolet, infrared, and radio: as with the pulsars discovered by the Mullard radio-telescope above. So my scheme for measuring brightness with ultimate accuracy might also give unexpected results and might trigger new interpretations. I am even more pleased that David Whiffen was able to ground my scheme firmly in the foundations of mathematical optics.

8

Rotating Things

One of the purest examples of an idea suddenly coming from the subconscious realm of the Random-Ideas Generator to the conscious mind was that of James Watt (see chapter 1). His idea for improving the steam engine came to him suddenly during a stroll on Glasgow Green in 1765; but it took him years of hard engineering effort to make it work. Furthermore, as Nicolas Carnot later noted (chapter 15), his cold external condenser is a direct counterpart to a steam engine's hot external boiler. Yet that inspiration of James Watt was crucially important. He is rightly known as the father of the steam engine. Arthur C. Clarke extols the steam engine, with the crank and valve gear on its rotating wheel, as the most visually appealing of rotating machines. In my TV career, I have also devised visually appealing rotating machines. Here are some of them.

Golden Syrup

I got the idea while spooning golden syrup. My RIG said, "Do this for Yorkshire TV!" The culinary "twiddle," or honey, spoon is well known. It is basically a narrow rod you can turn. You can hold lots of honey, or jam, or syrup, on it. By contrast, a static spoon merely holds the liquid in its bowl; the rest drains off. All good cooks and painters have to develop the right sort of turn, and so did I. H. K. Moffatt even has a paper on the physics of twiddle spoons![1] He defines a speed that gives a fairly even distribution of viscous fluid. At lower speeds, asymmetric lobar deformations occur, like backward-breaking waves. At higher speeds, disc-like instabilities develop, and the viscous liquid may be thrown off. I was glad to have come across

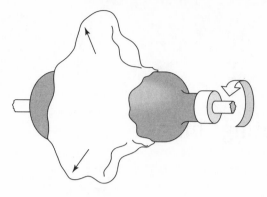

FIGURE 8.1   Syrup on a Spinning Bottle
Sticky golden syrup can be held on a spinning bottle. Too slow a spin just lets it slump off. Too fast a spin forms an irregular disc. The syrup streams away from the bottle (arrowed) and may fly off.

that paper; it gave me a sort of scientific anchor in my approaches to YTV. I imagined an apparatus in which you could spin a rotating glass tube and pour viscous golden syrup onto it. In the event, I used a soda bottle as my rotor; this was safer, cheaper, available in many identical copies and easier to transport from Newcastle to the YTV studios in Leeds.

I built my apparatus from neatly covered planks of wooden board. I powered it from an electric drill, its speed governed by a variable transformer. A little pulley on the drill drove a big one on the shaft with the bottle. The camera and the audience saw that turning bottle. Incidentally, I have always liked a studio audience. It acts as a sort of guarantee that what the camera is seeing is real. The claimed demonstration is actually happening in the studio and is not just a piece of clever editing.

The show's physical sciences moderator Magnus Pyke had several jugs of golden syrup and controlled the knob on the variable transformer that determined the rate of spin of the bottle. A large photographic developing tray, under the rotor, collected the syrup that fell off. As usual, the whole apparatus had to come apart easily. It let us conduct rehearsals and explore trial takes. When a bottle was covered in syrup, we could remove it and put on a new one. (I had brought lots of bottles with me.)

Everything worked. Magnus could pour golden syrup onto the rotating bottle, vary its speed of rotation, and expound on the various effects.

The whole thing even got into a promo for the show. Said Magnus in that promo, "in which I get into a pretty sticky situation."

Moffatt's paper seemed sound. If the bottle turned too slowly, the syrup simply slumped sideways off it into the tray. There was a speed at which a lot of syrup could be held about stably on the bottle. But too fast a spin was even more fun. The syrup formed disc-like instabilities that soon got asymmetric (fig. 8.1). Ultimately the rig, racing at high speeds, lobbed blobs of syrup at the TV cameras. My sense was that the bottle held the maximum amount of golden syrup at about one-third of the highest feasible speed. Magnus did not go into the physics, but I like to imagine that my demonstration had a nationwide impact—a sudden demand for twiddle spoons.

Rolling Cans: The Drag of Syrup

We faced a continuing problem in the YTV science office. We wanted to interest the audience in science, but we also wanted to show them things they wondered about in their domestic life. This steady challenge kept my RIG busy all the time. But I once thought of a very common predicament, maybe with a scientific answer. Suppose you have a can of food, but the label has come off. What is in the can?

Given the invariant volume of a label-less can, one trick would be simply to weigh it. A heavy one would have a heavier, denser food inside it. But weighing a can is rather boring TV. Furthermore, canned foodstuffs are essentially all water, so that all the weights would be much the same. Another trick was to put the can in the freezer. If, on taking it out, it warmed up rapidly, it had to contain some foodstuff of low thermal capacity. This was more surprising than weighing; but again most tins contained mostly water, and they all warmed up at much the same rate.

Faced with a label-less can, most people would shake it, and hope to judge the contents from the noise. This got me thinking about the mechanics of a can and how to study it. The simplest approach was to roll it down a shallow slope. This could be very good TV. Furthermore, there was some cunning science in it. Consider a can with a very liquid content (strawberries in light syrup, say). Allowed to roll, it accelerates rapidly; the can rotates but the fluid contents do not, or not much. So in a race against a can with a pretty solid content, say baked beans with sausage, it takes off fast and is soon winning. The baked beans with sausage starts slowly, as its whole contents have to rotate. But once it is going, it is free of lossy fluid friction inside, which slows the can containing strawberries in syrup. So the baked beans with sausage catches up and overtakes the berries in the end.

I spent a few happy days buying all sorts of cans of food and studying them. I went for the standard large cans. In the United Kingdom, they are commonly known as "tins," as the steel sheet from which they are made is tin plated.

Technology has advanced in recent years. Many cans nowadays are made from a "tumbler" drawn in one piece from bulk steel. But my old cans were fabricated from sheet; they were cylindrical, with identical hooplike ends. They would truly roll.

The producers and I loved the idea of holding a can race, with Magnus Pyke commenting on it excitedly as if it were a horse race. But I wanted a large plank for my race, maybe 3 meters long and wide enough for the cans to run free without hitting each other or falling off. I could not get such a thing in my Mini car. YTV claimed, of course, to have some good equivalents in its studio stock.

I did not believe this claim but had no choice. I took a load of food cans down to Leeds and, sure enough, YTV offered me two uselessly short planks for the item. We finally got the demonstration camera-worthy by doing it on a long bench, one end of which was elevated a bit by a piece of wood. The resulting can race was truly worth watching!

The Artificial Tornado

The artificial tornado did not spin syrup. It spun air. Rather foolishly I suggested the idea of the tornado to YTV, and the producer said, "Do it next week!" I had no time to optimize the design. Nor could I test it except in the studio. So it was a strong trial of my physical intuition (see chapter 4). I just had to build the thing as I imagined it and hope that it would work.

I planned it as a sort of open telephone booth with a suction fan at the top. I built its frame from Speedframe hollow steel sections, which can be assembled neatly and strongly with plastic inserts and dismantled again with a mallet. At the top went the fan and fan motor, mounted on shelving girders. There was no time to be clever or subtle. I borrowed an electric motor, of about 0.4 horsepower, from the Newcastle chemistry department engineers, bless them.

My next problem was the fan it had to drive. I went to Minories, a local car parts shop, to get a car cooling fan. I waited in line there with a lot of mechanics. When I reached the counter the bored assistant said, as he had done dozens of times that morning, "What make? What model?" "It doesn't matter." I replied. "I want a cooling fan that will fit in this trash can" (I had brought one along; the design in my head used a fan in a section of trash can). "If it works, you'll see it on TV this Thursday!" The assistant was captivated and brought out a lot of fans. I finally chose one from a "Stag" van; it had a strong polymeric fan that seemed to fit my trash can. Meanwhile people lined up behind me muttered as they clutched broken crankshafts and blocked carburetors.

Then I assembled the whole thing. I bought other parts locally as I needed them or exploited junk I had about. When the tornado was ready, I turned it on. My motor had a no-load speed of 2,850 revolutions per minute; with the fan, it went at 800. The whole thing got very hot. No matter, I thought; (a) it won't be on for very long and (b) it is ferociously air-cooled. In fact it is the most air-cooled motor there has ever been.

Wrong! The motor coils were shielded from the cooling draft by magnetic and constructional metal. They got much too hot much too fast for any demonstration and rapidly began to smell. I dashed back to the chemistry department and managed at short notice to borrow a 0.25-horsepower motor designed for 1,425 revolutions per minute. It was attached to a vacuum pump and was possibly faulty. It was the motor that was faulty, not (as I had hoped) the pump. I managed to correct it and put it on the tornado. It had the same mountings as the previous one and went on easily. And it did not get worryingly hot. In fact the new motor spun the cooling fan rather well.

Then I had to push my physical intuition even further. A tornado is not a mere updraft. It is an updraft with converging winds. Any tornado, or hurricane, or circular storm takes place on a rotating Earth. It sucks in rotating winds that converge on it, counterclockwise in the northern hemisphere, clockwise in the southern one. By the law of conservation of angular momentum, those winds intensify as they approach the updraft "eye." I planned to make the swirl around the tornado with four vacuum cleaners, blowing. But I could not try the monster in my house. I would just have to take the whole thing down to the Leeds studio and hope. I vaguely planned to space the vacuum cleaners widely apart and let their draft be intensified by the inward suction.

When I got to Leeds, YTV provided me with space, four vacuum cleaners, and lots of confetti. I put that confetti in the middle of the tornado, under the updraft fan. My idea was that when I turned on the updraft by itself, nothing much would happen. When the vacuum cleaners gave that swirl, the updraft fan would suck up the confetti and blow it all over the studio.

Amazingly, it all worked! When Magnus turned on the vacuum cleaners, my rig blew confetti all over the studio. Contrary to my notions, the tornado seemed most dramatic with the vacuum cleaners quite close in. It was the centerpiece of the show! But after that triumph, YTV and its actors cursed me bitterly. For some of the confetti landed on the studio lights. And mere days later, that studio was being used for some other purpose, maybe a costume drama. Whenever a lighting technician adjusted a light from the console, the disturbance might dislodge a piece of confetti. It would come zig-zagging down in front of the camera and ruin the shot. They would have to shoot it again!

Later I used my artificial tornado again, on German TV, in a Westdeutscher Rundfunk studio in Cologne. It is the things you don't know that give you most trouble. Far too late, I found out that I should have brought some British confetti with me. German confetti is made of cardboard and thrown like buckshot, and no amount of draft will lift it. I and some high-ranking WDR officials spent an enlightening hour or two tearing up Kleenex tissues to make a pile of paper that my tornado could lift.

A Paper Saw

The ideal scientific TV demonstration uses something that we all know, but explains it in a new and scientific way. One of my triumphs was a rotating paper saw. Many of us have cut ourselves on a moving paper edge (I certainly have). But how to do it on TV?

A spinning paper saw, said my RIG. What domestic object spins fast? A kitchen mixer was far too slow; so was an electric drill. Ultimately I settled on a coffee grinder. If I removed the grinding mechanism and made an extension to hold a paper disc, it all worked well. The disc only had to be a few centimeters across, for the grinder spun it at some 18,000 revolutions a minute. Centrifugal force pulled it rigid. It was far too ferocious for Magnus just to wave around. I made a mechanical mount for it, that of an electric drill. With the handle, Magnus could bring it down on whatever he was going to cut. If the saw went wrong, or for any other reason he let the handle go, everything would snap back safely.

What should I cut with my paper saw? A metal bar was too tough. Oddly, a wooden pencil could be cut just halfway. This showed me what was going on. The thing being cut was being frictionally heated, burned through by the racing air-cooled paper disc. A pencil with a graphite rod in the middle lubricates that friction. A solid wooden dowel seemed ideal but would mean nothing on TV. So I used a small wooden-handled dishwashing mop, which most TV viewers would instantly recognize. Magnus could cut the end of the mop off with the paper saw and show what he had done. Furthermore, he could then show the sawn part to the camera.

Even better, Magnus could cut the paper disc on camera and with scissors, showing that it *was* paper. If it was not quite round; the extended bits were soon machined off as it sawed. Indeed, the disc shrank in use, and I had to keep changing it in the rehearsals.

I never discovered a truly reliable paper. Brown parcel paper seemed good; so did magazine paper (which has various particulate loadings to help color printing). But any sort of paper might tear unpredictably. I liked the idea of using a £10 note. Cutting a disc out of a bank note would make gripping TV, and the Bank of England would replace it if we recorded its number. Later I took the equipment to Cologne to Westdeutcher Roadfunk TV. The Germans were horrified at the prospect of cutting up a sacred Deutschmark note on TV. I had to explain to the WDR team that in Britain the currency is less highly regarded.

Watching TV in a New Way

As a science consultant to YTV, my mandate was to make science real to the viewers. Adam Hart-Davis, then part of the same team, once said: "We know one thing about our audience. They have all got a television set." So I began to muse on ways of bringing TV technology home to its users.

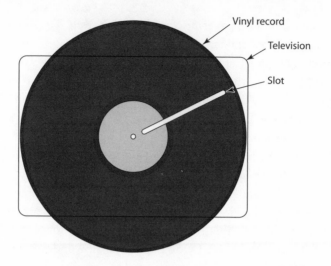

FIGURE 8.2   Showing TV Line Speed with a Slotted Vinyl Record
A TV picture is simply a line running down its face, 60 times a second in
the United States (50 in Europe). If a vinyl record, bearing a cut slot, is spun
clockwise in front of the television screen at this speed, the slot will travel at the
same speed as the line, and much of the right-hand side of the TV image will be
visible. It works!

TV, of course, is a line moving down a screen, 50 times a second in
Europe (60 in the United States). At that high speed, it seems a whole
picture to the human eye. I began to imagine a disc with a slot cut in it,
spinning at that rate in front of the screen. On one side of the disc, the slot
would be going down. It would be moving with the line and would show
you a lot of the picture. I agreed to try to make such a disc.

What to make it out of? Ideally, I wanted something familiar to a TV
audience. My physical intuition felt quite good about using the old vinyl
records. Years ago in a failed project, I had made noncircular holes in such
discs and had cut them up. In those days, of course, any member of a TV
audience would recognize a vinyl record. I felt that a slotted vinyl disc
could be made to run at 50 revolutions a second and could be looked at
by a camera synchronized and in sympathy with a TV picture behind it
(fig. 8.2). When the two were not perfectly synchronized, it would show
an interesting interference pattern between the two.

I had a lot of trouble building that demonstration. It is not easy to cut a slot in a standard long-playing record. I used my old lathe as a milling machine and took everything slowly. Then I had to spin the record at 50 revolutions a second in front of a TV screen. It had to be well mounted on a good bearing. To work well, without vibrating, the slotted disc had to be in perfect balance—and more by good luck than good judgment, it was.

I wanted my rig to be as transparent as possible, so as to show the TV screen behind it. So I mounted my slotted record on a transparent plastic strut and drove it via a rubber belt from a variable-speed motor at the bottom. Then I assembled the whole thing and edged it gradually up to speed. On the screen I saw lots of fascinating interference patterns and finally quite a decent fragment of picture. It worked! Even so, a big object spinning fast in a TV studio worried me greatly. If the disc broke under centrifugal force, it would throw sharp-edged plastic shards all over the place, maybe causing injury as well as wrecking the item. But everything stayed together. Even better, some of the TV audience may have learned something. They may have realized that they were just watching a moving line. I never found out what rotational speed would break the record, and so I never knew my margin of safety. The disc was far too precious to sacrifice!

9

Explosions and Fuses

Like many a future chemist, I played with pyrotechnics in my boyhood. One such chemist has even put the experience into verse.[1]

Much later, some of my hard-won insight came in handy for lectures and TV. In retrospect, my experience contrasts strongly with other parts of this book. I learned pyrotechnics over years, but used it as an adult in very swift bursts, in lectures and on television. What I learned was essentially emotional—I was not frightened by the business. I could play with it. All this is relative, of course. My friend Fred Peacock was horrified to learn that I showed acetylene explosions in lectures and on TV—to him the gas was very dangerous. Conversely, I was horrified by some of Brian Shaw's exploits in his lecture on explosives. Experiments that were simple to him really worried me. Yet my boyhood experience and relative lack of fear often paid dividends. Several times I gave exciting lectures and sometimes got pyrotechnics on TV. To most of the TV crew, not to mention the audience, these demonstrations were very frightening. With my calmness, experience, and knowledge, I handled them with panache.

Gunpowder and Other Solids

About 1950 my creative friend David Andrews introduced me to his ingenious explosive, "chlorocellulose." He made it by soaking cotton wool in a solution of sodium chlorate and then drying it. In those days you could buy sodium chlorate in cans, as a weed killer. David Andrews, and doubtless many other budding chemists, mixed it into fireworks. After he invented chlorocellulose, he went on to make mixtures of sodium chlorate with sulfur and charcoal, both as an explosive and as a rocket fuel. He explored the chemistry carefully, and his best explosive mixture was "Fuel 14."

I was interested in how long an ignition lasted. I made a long thin trail of a mixture, lit one end, and saw how it burned. Commercial gunpowder always burned faster than anything that I or David Andrews could make. I read somewhere that gunpowder burns by the diffusion of hot sulfur vapor. This seemed to fit the few facts I knew.

The ingredients of gunpowder are potassium nitrate, charcoal, and sulfur. I developed a great respect for it. No matter how exactly I proportioned the ingredients, nor how finely I ground them, nor how carefully I mixed them, the powder I got burned much more slowly than commercial gunpowder taken from fireworks. I have read that commercial gunpowder is wetted, and the damp mass is "corned," or ground further, before being dried. The resulting coarse powder consists of grains, each of which is a fine mixture. To rival it as an explosive, I had to replace its potassium nitrate by sodium chlorate weed killer. I suspect that sodium chlorate gives out heat as it decomposes and this helps an explosion along. David Andrews's Fuel 14 had the same sort of composition and was probably as good an explosive as gunpowder. It was not as safe, though; it went off violently if you hit it.

Both David Andrews and I built rockets. Their pyrotechnic contents had to burn for several seconds. So we rammed our rocket fuel in hard. I reasoned that this reduced the crevices through which vapor could diffuse and slowed combustion. One design we explored had a central hole in the rammed solid (we made it by pushing a dowel up the nozzle; later we carefully withdrew it). The pyrotechnic material burned mainly at the surface of the hole, so that its combustion spread radially outward. Modern solid-fuel rockets use the same trick.

Once I had a nasty surprise. I wanted to make a truly solid rocket fuel. I thought of soaking my powder in a solution of polystyrene (the solid packing material Styrofoam) in dichlormethane. The polymer solution was a viscous liquid; mixed with rocket fuel it gave a sticky goo that I packed into a rocket case. I reckoned that it would dry to a solid with no holes in it at all. It should then burn at a slow, stable rate. I failed to appreciate the chemistry. Dichlormethane dissolves sulfur. As it evaporates, it deposits the sulfur in a highly divided form, and this, perhaps, oxidizes in the air to sulfuric acid. The strong acid can react with the sodium chlorate oxidizer and can set it off. Anyway, my clever new rocket fuel was spontaneously inflammable. Days later it went off and blew the door off my bedroom closet.

Much later, that fearful spontaneously inflammable rocket fuel came in handy, in a Yorkshire Television Ltd. demonstration. Magnus Pyke was showing a crazy experiment of Sir Humphry Davy's. A red-hot cannon ball, focused by two paraboloidal mirrors, concentrated its image on something Magnus held at the focus in a pair of tongs. That something was a piece of paper based on my rocket fuel. Throughout the 2.5-hour car journey from Newcastle to Leeds, I was worried that it might go off spontaneously, before even getting hot. But it worked. YTV got their shot.

And why does a rocket have a stick? Why does it have to lift that dead weight? Our guess was that it stabilized the rocket in flight. A stick might give the rocket such a large moment of inertia in the direction of flight, that is, if it turned at all, it would turn slowly. (In that case, a stick pointing ahead of the rocket would also work. We never tried it.) Second, the stick keeps the center of gravity of the rocket behind the nozzle. Later I discovered that the great American rocket pioneer Robert Goddard had once built a rocket with a nozzle at the front. He had clearly pondered the same argument.

I once tried to make a rocket without a stick. I knew nothing of Goddard at the time, but I recall my reasoning. A solid-fuel rocket is a charge of a combustible pyrotechnic fuel in a cylindrical case. The case is closed at the top but has an opening at the bottom through which you light the thing and from which flame and hot gas emerge as the pyrotechnic material inside burns. The resulting jet of hot gas exerts thrust on the cylindrical rocket body; with luck it takes off. The exit nozzles of the rockets that I and David Andrews made were generally cardboard. They (a) usually widened and lost shape as the flame and gas roared through them and (b) contributed nothing to the stability of the rocket—indeed they could push it off course if the nozzle happened to widen more on one side. Hence that stabilizing stick. My brilliant stroke (as I thought) was to make the nozzle of Pyruma fire cement and to shape it to spin the gas as it emerged. This should spin the rocket the other way. Stabilized by that spin, the rocket would not need a stick. Cunningly, I planned to shape the fire cement

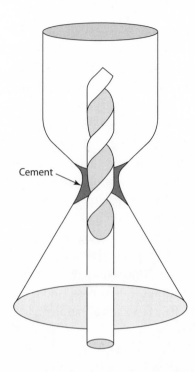

Cement

FIGURE 9.1   Rocket Nozzle from Fire Cement and a Twist Drill Bit

The rocket itself, containing some pyrotechnic mixture, is a cardboard cylinder above the nozzle. Only part of it appears in the drawing. The nozzle region at the bottom is conical and leads to a region whose fire cement surrounds a twist drill bit. I planned to twist the bit out when the cement was nearly set, leaving a shaped rocket nozzle of hardened fire cement. Through this fluted hole, the gas and flame of the rocket would emerge spinning. That spinning efflux would expand in the lower conical part of the nozzle. My hope was not only that it would drive the rocket into the air. By spinning it the other way, I hoped it would give it rotational stability. It did not work (like so many of my clever ideas).

nozzle with a twist drill bit (fig. 9.1), which I could twist out when the cement was nearly hard. Then I would let the nozzle set completely.

Sadly, my clever idea (like many other of my clever ideas) failed in practice. My fire cement nozzle gave little spin to the rocket and seemed no better than usual.

David Andrews continued to develop his simple "choked bore" nozzle. He soaked a cardboard tube in water to make it soft, throttled it to small diameter by tightening a string around it, and allowed it to dry and harden. His rockets were as good as mine and maybe better. And we continued to use sticks. My favored stick was a long, light stalk from the goldenrod plant, and I implicitly accepted the center of gravity theory of its action. I balanced the whole rocket on my finger to check that the center of gravity was indeed behind the nozzle. Then David Andrews invented the hollow-tube rocket. His insight was that the faster a rocket burns its fuel, the more efficient it is. So his new rocket burned its fuel almost explosively, in a small fraction of a second. He also reasoned that it would help if on launch the rocket blew something out the back, instead of relying on its own gas. His hollow-tube rocket was about 30 centimeters long and 13 millimeters in diameter. It was made of a few turns of brown paper, glued together and tied at the nose. Its motor fitted inside that nose, and was lit by a little side "nose fuse." The blast of its gas came down its hollow paper tube. The new rocket was launched from an aluminum-alloy tube about 12 millimeters in diameter and maybe 70 centimeters long. David pushed this a few centimeters into the ground, slid the hollow-tube rocket all the way onto the long projection of the other end, and lit the fuse. As David remarked, a rocket launched from that tube, "Blew the Earth out the back." That launching tube also guided and directed the new rocket in its take-off. Its motor burned so quickly and violently that it had used up all its fuel by the time it had flown about a meter from its launching tube. But by that time it was going at about 70 meters a second. After that initial blast, the Andrews rocket flew through the air as a passive ballistic object. It was not disturbed by the thrust of its long-dead motor. With the weight of that motor in its nose, and three little cardboard stabilizing fins at its rear, it flew fast and very straight, like a dart. It was a brilliant invention.

As usual, after David had invented his new rocket, I explored some improvements. I improved his fast-burning motor. I tried to make the whole thing bigger and failed: a paper tube 13 millimeters across can withstand sudden internal pressures that may burst a 2 centimeter or 3 centimeter tube of the same construction. I tried a conical form; it worked, but was much harder to make. I perfected a clever improvement to the Andrews rocket by elaborating it into a two-stage rocket. The second stage blew the first stage out the back, a neat trick that NASA has not taken up.

David and I faced a crucial problem with rockets: once we got a rocket up, it had to do something visible at its maximum height. David had invented the inflammable "doughball," a mixture of flour and sodium chlorate, wetted to a paste and then allowed to dry. It burned for several seconds as a bright yellow star. A group of them could be flung out of a rocket by a charge of Fuel 14, lit by a fuse. This both lit and dispersed them. Later I invented a device that in principle set off a rocket payload at its extreme height. Even if it failed, the payload would ignite some time. (If anything went off in the air, honor was satisfied.) We once had a scheme to send a rocket so high up that if I launched it in Orpington, David Andrews could see the payload ignite in the sky in Bexley—they are about 10 kilometers apart. We never did it; it would have challenged the telecommunications of the day anyway. Nowadays, the mobile phone would make it easier.

We did better with "bangs," making only "low" explosives with mixed ingredients. Almost any ignitable mixture in a small paper case and set off by a fuse will burst that case with a bang. Our biggest fireworks were quite small—about 20 milliliters of volume. We never made anything truly vast and frightening. But we still argued about their ideal chemistry and designed them carefully. Thus each serious explosive object had a vigorous explosive core that set off the main charge.

I am now somewhat ashamed of the low explosives that I and David Andrews made; real pyrotechnic chemistry is far more advanced and professional. But we asked some of the right questions, I think.

We often built these fireworks into underwater "mines." These were fireworks designed to explode under water. An underwater mine needs a delaying fuse to set it off, and we used Jetex Igniter wick (see below) in a waxed-paper drinking straw. Once the flame of the wick was inside the straw, you could safely drop the mine in water. The burning Igniter wick generated a rapid string of smoky bubbles from the end of the straw and stopped water getting down the straw and into the mine. The fuse would burn down until it set off the main charge.

One crucial invention I made as a teenager was wax waterproofing for our mines. It remained in my mind as a simple technique, clever and effective. I have since used it several times on TV, in some spectacular underwater fire or explosion demonstrations. As a teenager I reckoned that nothing would explode if merely heated to the boiling point of water. Wax melts below that temperature. So I melted bits of household candle in a can in a small saucepan containing boiling water. I could then dip the mine in the melted wax. The result was a waterproof mine. It usually needed a weight to sink it, which I attached to the string of its construction. When David Andrews and I exploded such a mine under water, we got a big splash, a little thumpy noise, but a sudden pulse of pressure that shook the ground. We relished that!

Once I tested a mine by lighting it and putting it in my mother's big zinc-plated steel-sheet tub that she used for rinsing clothes outdoors. I had filled the tub with tap water. The mine went off without much bang, but the hydraulic pulse split part of the seam of the tub. Later I tried to solder that seam together again, but I fear that the tub was never reliable afterward.

Yet that creative experiment stamped an idea in my mind. Years later, when I was expounding on the torpedo on German TV, I fired a gunpowder charge under water. It was held against a rectangular tin can containing air. Using that clever waterproofing technique I had invented as a teenager, I waterproofed that "mine" or "torpedo" with wax. The exploding gunpowder, with its sudden pulse of hydraulic pressure, crumpled the can most dramatically. In air, without that hydraulic effect, the same charge made a much louder bang, but merely knocked the can over.

Fun with Gaseous Explosions

The gas explosion, of course, is a crucial part of modern technology. Every gasoline engine is driven by the gas explosions in its cylinders—though jet engines and gas turbines are internal-combustion engines powered by continuous flames.

The gas explosion has one great advantage over the solid charge—its energy is much lower. You can drive an internal-combustion engine safely with gas explosions. Most engines have a capacity (the summed size of all the cylinders) of a liter or so. Indeed, you can show a gas explosion in a lecture or on TV in a tin can containing a fraction of a liter. A solid charge filling a tin can would kill you—it would be like a hand grenade. And in most gas ignitions, the flame travels very slowly: a meter per second or less. I have shown such a flame moving through a mixture of gas and air in a lecture, visibly and slowly. Indeed, maybe only two gases give a flame that propagates fast enough for a lecture explosion—hydrogen and acetylene.

David Andrews introduced me to acetylene. It has two advantages over hydrogen. Firstly, its mixture with air ignites in concentrations from 3% to 80%, so you need no precision to make a bang. Secondly, you can make it by putting a chunk of calcium carbide in water. As a teenager I used to make a hole in the side of a lidded Nescafé can (the small size, about 350 ml), put some water into it and push the lid on, stand the tin on the ground with the lid at the bottom, push a bit of calcium carbide into the hole, and then apply a light. I got a mighty bang, and the tin was blown into the air. Its content of water went everywhere, so I got wet too.

I now suspect that the mass of that water helped to accelerate the can upward. That was David Andrews's recipe. Later I got cleverer at making acetylene explosions.

The biggest I ever made featured about 40 liters of acetylene and air mixture in a plastic carton. The loudest had 3 liters of it, with an acetylene and oxygen mixture in a big Coke bottle. I got my nephew, Tommy, to press the button. The bottle went off with a frightful bang and set off a car alarm 50 meters away. I also developed a lecture demonstration. It used a gas explosion to throw a beer can at 60 meters a second, faster than any Newcastle thug could throw it. After showing a gas explosion in a lecture, I would tell the audience: "When you are traveling fast down the motorway, what you have just seen happens 200 times a second in the car engine. And that's what drives the car!"

A very simple gas-explosion demonstration uses a lidded tin can with two holes: one in the lid and one at the bottom end of the can. You fill the can with gas (old-fashioned domestic coal gas used to be the combustible gas of choice; nowadays you use hydrogen or modern North Sea domestic methane gas). You light the gas at the top and get a flame from the hole. This depends on the gas being lighter than air, which coal gas, hydrogen, and methane all are. As gas burns at the top hole, air enters at the bottom. The gas-air mixture inside the can approaches explosive composition; the flame coming out of the can gets smaller and more intense. Ultimately it burns down into the can faster than the mixture can come up through the hole. Flame spreads inside the can, which suddenly explodes. I always have the can inverted, so that the can itself, and not a mere lid, is blown into the air (fig. 9.2).

I was once watching this demonstration, and the lecturer had set up two cans. My Observer-Reasoner noticed, and my unconscious mind was greatly intrigued, when both exploded together. Why? Musing about it later, I reckoned that the sound-pulse of the first explosion pushed in the flame of the second, and set it off. Here was a possible TV demonstration, if I could get it right. I played around with it, finding out how close the flaming cans had to be for one to be fired by another. One or two meters seemed to do it, and hydrogen seemed better than methane. Later I set up for German TV a demonstration with three syrup tins of about 350 milliliter capacity. They were filled with hydrogen and their lids sealed on with nail polish (hydrogen diffuses rapidly through even the small-

FIGURE 9.2 Hydrogen-Filled Cans Exploding Upward Together

Three inverted cans filled with hydrogen explode simultane-ously. The first to explode sets off the others. The flying cans leave their lids behind; these can be seen on the bench. A can exploding upward goes too fast for TV, so I attached a Christ-mas garland to each one. By suddenly flying into the air, the garlands show how fast the cans explode upward. In figure 9.3, I show how a similar garland can make a rocket-bottle visible to a TV camera.

est hole). To make each can more visible to the TV camera I attached a Christmas garland to it. They were intended to explode together. They did, too, with the audience present—though I never made the arrangement very reliable.

Over the years I have simplified my gas-explosion technology and can now deploy it routinely. As so often, the basic observation was an accident. My friend Mike Alder once dropped a bottle of fizzy Coca-Cola in a supermarket. The screw-cap came off, and the bottle roared away, blasting Coke back out of its neck and spraying supermarket customers with the fizzy beverage. Mike acted innocent and pretended to have no connection with the incident. But when he told me about it later, we began to play with soda bottles. We discovered that they are very tough toward internal gas explosions—as indeed you would expect for commercial containers of a gas-pressurized beverage.

Even a slow-burning gas like butane could blast the water out of its neck, when the bottle zooms off like a rocket. A fast-burning gas like acetylene is even simpler. A mixture of acetylene and air, in a Coke bottle, can be lit at the neck. The burning gas rushes out of the bottle, which zooms off at great speed with a satisfying roar. Again, to make it more visible to a television camera, I could tie a light long feathery Christmas garland around that neck (fig. 9.3).

I showed this splendidly once, on German TV in a Christmas show. The studio had a tree, which was hung with many tinselly decorations. My rocket had its Christmas garland around its neck. I fired it very inaccurately, as I often did. It took off with a mighty roar and went straight into the Christmas tree. The garland wound around a branch, and there the bottle hung, looking like a crazy new decoration. The cameraman, bless him, had followed the whole action. It was a lovely shot.

One trouble with gas explosions and flames is that they are hard to see—especially in a brightly lit lecture theater or TV studio. So I have often wanted to brighten a gas flame. I once managed it with sodium bicarbonate powder, making the flame yellow. But it was hard to mix powder and gas. The best trick I found was to mix the gas with a little ethyl borate. This volatile liquid boils easily; its vapor mixes readily with the gas and makes the flame green. A camera or an audience can then see it.

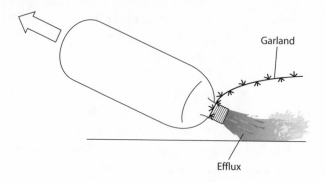

Garland

Efflux

FIGURE 9.3   Two-Liter Fizzy-Drink Bottle as a Rocket
I used a 2-liter bottle from a carbonated beverage such as Coke. Indeed, an accident with fizzy Coke started my research. Later, I used water expelled by a gas explosion. Later still, I got the ignition efflux from an acetylene-air explosion to work as well. The gas mixture in the bottle is lit at the neck. It explodes, rushes out, and blasts the bottle forward as a rocket. For better visual appeal on TV, a long light garland can be attached to the neck of the bottle.

Another chemical gas reaction I have played with is the ignition of hydrogen in chlorine. This too goes at an explosive speed. It can be set off by sudden bright light, as made by a photographic flash unit. Hydrogen and chlorine are both available in cylinders, but I am wary of them. I preferred to make my mixture by passing electricity through hydrochloric acid, using graphite electrodes. You get hydrogen and chlorine in equal quantities. I had a lot of trouble with this reaction—the chlorine attacks almost everything, and it is very hard to make the electrolytic cell. And you need a rectifier—you have to use direct current. I used an old-fashioned transformer and rectifier combination. It got hot, but drove my electrolytic cell very convincingly.

Once I had an arrangement that seemed to work, I played around with the gas mixture it gave. Sudden sunlight could set it off—I had a nasty surprise this way—but steady electric lights seemed not to. So I could show it in a steadily lit lecture theater or TV studio. I only needed a few hundred milliliters of the mixture in a plastic bag to make a fine explosion (fig. 9.4).

This made a splendid lecture item, featuring both chemical and electrical equipment. But I dislike chlorine. It is a poison gas, and even a trace

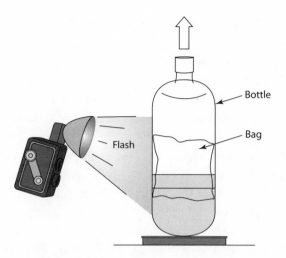

Bottle

Bag

Flash

FIGURE 9.4   A Hydrogen-Chlorine Explosion in a Two-Part Bottle

I cut a 2-liter Coke bottle in two and fasten the bottom part to a baseboard. I can then push the top half, with its screw cap firmly in place, onto the bottom half. I make a mixture of hydrogen and chlorine electrolytically (this is a tricky bit of electrochemistry, fascinating to an audience or the TV watchers by itself), accumulating the mixture in a polyethylene bag. I pop the bag into the bottle, and push its two pieces together. I then tell the watchers that the mixture I have made is very unstable. It does not quite explode if you look at it, but it explodes if you take a photograph of it. I then flash it with my camera; the gas mixture explodes and fires the top half of the bottle up, to hit the ceiling. Few people know that the hydrogen-chlorine gas mixture explodes in sudden light.

smells awful. In the apparatus as I developed it, I bubbled the electrolytic gas mixture through water to remove suspended drops of acid and then passed it through calcium chloride pellets to dry it. As the rig evolved the gas mixture, I accumulated it in a little plastic bag. When the bag was nearly full (it could hold a few hundred milliliters) I put it in a Coke bottle that I had cut in two. I had screwed the bottom to a baseboard and could push the two parts together with the bag inside.

I could then tell the lecture or TV audience that the gas mixture I had made is extremely unstable. It does not quite explode if you look at it, but it explodes if you take a photograph of it. Then I took my flash picture of it, when wham! it exploded. It blew the top part of the bottle at the ceiling. I suspect that this explosion is the fastest possible reaction and goes at the speed of light. A flash traversing the mixture sets it off; it emits more light as it explodes and propagates at that speed.

These days, much combustible liquefied petroleum gas (LPG) gas is sold in canisters. It is liquefied by its own pressure. Propane and butane come like this and in Britain so does the mixture Calor gas. You can often hear the liquid sloshing around inside the canister. The volatilized gas comes out through a simple tap arrangement on the canister. I have played with it many times, but its flame propagates much too slowly to make an explosion. Instead, I can let it out through a fine hypodermic syringe needle. The jet of unburned gas travels through the air. It expands and slows as it comes out and can be lit a few centimeters away in free air as a "detached flame." I once planned to make a big detached flame for a chemical exhibition. Sadly, the scheme came to nothing, and it remains one of my many fantasies.

The Metal in the Middle

That popular handheld firework the sparkler is made by dipping a metal wire in a gunpowdery paste containing lots of iron filings and letting it dry out. Light it at the end, and a zone of combustion passes slowly along it. The central wire controls the rate of burning by conducting its heat along. That heat comes from within. It throws all the products of combustion outward. They include the heated iron filings, which burn in the air and give the sparks. Sometimes they even melt and divide; when the spark branches prettily. Yet those visible sparks are still very small. If they hit your hand, they don't hurt.

I have tried burning sparklers in pure nitrogen. They continue to burn, in the sense that a zone of ignition continues to pass down the wire. But they seem not to throw out sparks. I guess that they still eject iron filings, but these fail to burn in the nitrogen. It would be easy to burn sparklers in other gases, but I have not tried it.

In the next chapter, I tell how Yorkshire Television phoned me up frantically one day. They wanted, in 24 hours, some rig for lighting a lot of candles on a cake very quickly. I foolishly said that I could do it. My immediate notion was to use a chemical fuse, with a metal in the middle. Hectic experiments soon convinced me that any standard chemical fuse

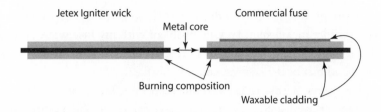

Figure 9.5

Jetex Igniter wick   Commercial fuse

Metal core

Burning composition

Waxable cladding

FIGURE 9.5 Metal-Cored Fuses

Metal-cored fuses burn at a steady rate defined by the heat traveling along the metal. Rates such as 1 centimeter per second, 10 centimeter per second, etc., are possible. Fuses with waxable cladding can even be made waterproof, using the clever technique I invented as a teenager. When covered with wax, such a fuse will burn under water. It emits smoky bubbles that make wonderful TV!

was far too slow for the job. And it would leave a metal wire behind. Yet I had said that I would do it!

My standard fuse was Jetex Igniter wick (fig. 9.5), made by the Jetex company for lighting its rocket motors. It burned reliably along long straws, through a heap of sand, though metal touch-holes and even through a hole drilled in thick metal. It was a brown threadlike product about 0.8 millimeters in diameter of some plastic stuff that I took to be an ignitable composition like the gun propellant cordite. It was formed around a central copper wire 0.19 millimeters across and burned at about 2 centimeters a second.

Later I realized that the metal core, like the central wire in a sparkler, controls the rate of burning of the fuse! Anyway, after that emergency with the candles, my RIG got musing about fuses. I made several with a metal wire in the middle, though I never made one as good as Jetex Igniter wick. One of my raw materials was cotton-covered electrical wire. I placed a sticky gunpowdery goo around it; this dried to a pyrotechnic composition. But ultimately I abandoned that cotton covering. I used thin lacquered copper wire about 0.15 millimeters across and wound sewing thread around it on my old lathe. I put a sticky goo around this core: a gunpowdery mixture made more coherent by adding water and PVA glue. My most successful coating gadget was a sawn-off hypodermic syringe barrel; I filled this with my goo and pushed the wire up through the Luer hole for the needle. If the goo did not stick well enough to the wire, I patted it with a nickel spatula. One problem with my fuse was that (unlike Jetex Igniter wick) it was stiff. Forced to bend, it bent sharply at an "elbow" and bits of the coating could fall off.

Later I came across thicker commercial fuse, for example, the type used for theatrical displays (fig. 9.5) and saw that it too had a central wire core. I bought commercial fuses with several rates of combustion (1 cm a second, 10 cm a second, etc.). These thick fuses had three layers. The middle was the metal wire that controlled the rate of burning; outside that was a layer of some gunpowdery composition that did the burning; outside that was a cylindrical fabric sleeve that held the whole thing together.

Now one's normal expectation is that water puts out fire. So I imagined a gripping TV demonstration in which this fuse would burn under water. I treated the fuse with wax, using that splendid waterproofing technique I had invented as a teenager. The resulting fuse burned under water splendidly, emitting sub-aqueous flame and smoky bubbles. I exploited it in two TV demonstrations, one for YTV and one for WDR in Cologne. For TV purposes, I needed to come up with a bit of terminal drama, which I provided by making the fuse light a simple commercial firework. Later I scaled up this demonstration and was able to light a volcano-type firework under water. Again, I first waterproofed it with wax. One unexpected bonus was the marvelous "gobbling" noise made by a firework burning under water.

10

Tricks with Optics

All modern visual technology depends on a simple optical illusion—that a rapid succession of visual "stills" gives the impression of a moving image. Many youngsters, including me, have explored this effect rather informally by drawing pictures on the corner of a book and then flicking through the book corner to see the moving image. Indeed, at the age of 10 or so, I graffitized the family telephone directories by drawing a simple pencil picture on the corner of each page, so as to draw a story.

The Moving Image

As so often, my creative friend David Andrews opened the field properly. He turned crude page flicking into a real mechanical art form. He drew a real cartoon film on a long paper strip. There was no projection system; you just looked at the paper tape as it jerked along. On the back of each frame he glued a cardboard strip for his "projector" to drive. Most commercial films use a film, a long sequence of transparent still frames held on a spool. It moves in front of a powerful lamp in a series of jerks (the screen rate, typically 24 frames per second). During its moments of stillness, the powerful lamp throws the image of the selected frame onto a viewing screen via a projection lens. That lens is adjusted to give the sharpest possible image; the whole thing is called a projector, and the audience sees the rapid sequence of projected stills as a moving image. David's machine was much simpler; the film was opaque, drawn or indeed painted on the paper of the tape, and you looked at it directly. Each picture was a little pen-and-ink sketch painted with watercolors, and the screen rate was about 12 frames a second. With his machine, each frame was still for half the time (during which the eye could see it) and was driven to the next frame for half the time (during which a black rotating shutter obscured the movement). I devised a slightly improved machine, in which the picture tape was still for three-quarters of the time and moved to the next frame in a quarter of the time. Both of us had to balance the need for optical continuity against the trouble of drawing hundreds of pictures. Each picture might only be 4.5 centimeters wide but still had to be carefully created. I drew mine with a magnifying glass and used a stencil to copy details from one frame to the next without too much jerking. My planned frame rate was about 11 frames a second. This gave a rather jerky but effective illusion of motion. The professional standard of 24 frames a second was adopted to let the film carry a soundtrack. European TV, which uses 25 frames a second, shows film a trifle fast. American TV, which delivers 30 frames per second, has a separate arrangement for transmitting film.

My major artistic achievement was to invent "endless films" drawn as a closed loop. You could then show a drawn film continuously. The enormous agony of drawing all those pictures was compensated by seeing each of them many times. I had a lot of trouble devising a connection that would go through my viewing machine. My loop masterpiece was the "Gunpowder Powered Internal Combustion Engine," 238 carefully drawn frames. Much later I transferred my animations to 16mm color film and thence to modern optical media. They still survive!

The crude cartoons that David and I drew transmitted visual information at about 1 megabyte per second (fig. 10.1). Professional 16mm film transmits maybe 20 MB a second, 35mm film about 100 MB a second, and high-class IMAX film some 1,700 MB a second.

And yet all this mighty optical machinery pays no attention to how our visual systems work. The eye does not perceive the whole scene. It concentrates on a tiny region that it scans around the scene. The brain then cobbles together a general visual impression. The human optic nerve is very narrow, and I guess that we can "see" only about 0.001 MB a second.

We now know that the eye scans around a scene in sudden movements called "saccades," during which the optical system is turned off. This brief "change blindness" lets you change the image. During a saccade, you can alter the color of a person's hair or even erase him from the

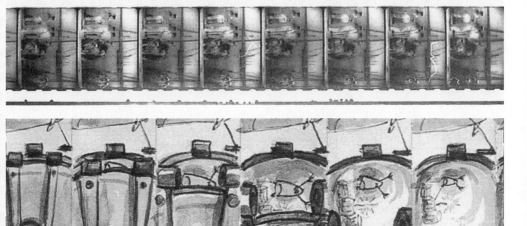

FIGURE 10.1   35mm Cine Film and Hand-drawn Cartoon Film
35mm cine film (*top*) is transparent and is projected onto a screen at 24 frames per second. The cartoon film (*bottom*) is of opaque paper and is viewed directly at 11 frames per second. Both exploit the illusion that a rapid succession of still pictures appears like motion.

scene. The viewer notices nothing. Saccades occur typically 5 to 50 times a second, so far unpredictably. So in 2001 Daedalus proposed a saccade-based video system.[1] He felt that the old framing technology should give way to a less well-defined one. He wanted the frames of a moving image to change only with the change blindness that saccades impose on the human visual system itself.

All the instincts of filmmakers have been that film is very expensive. Almost instinctively, they chose the slowest frame-rate and the narrowest film they could. By contrast, modern digital methods are very cheap and very sensitive. A modern digital system could have a very high frame-rate—say 300 frames a second. Daedalus's idea was that you would re-cord all those frames but only show a subset of them to the viewer. The system would choose a still image and hold it to the next saccade. Then, during the viewer's moment of change blindness, you would update it to the best next still.

The whole system implies a single viewer—as in modern TV or DVD technology. An infrared camera looking at the viewer's eyes would spot the saccades.[2] TV transmission, still bound by its 25 or 30 frames a second, would send every twelfth or tenth frame. As far I know, nobody is studying such a system, but my RIG feels good about it.

Making Candles for Television

A candle is chemically quite cunning. The solid wax melts to a puddle of liquid; this climbs the wick by capillary attraction and burns at the top, giving a luminous flame and melting more wax. When Yorkshire Television Ltd. proposed doing an experiment with candles, I bought lots of domestic objects for the show. I stuck wicks in them and tested them for candle and TV potential.

I had a lot of fun making new candles. I made butter and margarine burn at a wick. I stuck a wick into several brands of shoe polish and into an opened can of sardines in olive oil. I tried to light a bar of soap at a wick and to turn oranges and lemons into candles. These failed on me, but I fared better with oily nuts. I lit a potato chip too; it acted as its own wick. For TV purposes, I made many general-purpose "wicks." Each was a candle wick stuck onto a needlepoint. The presenter could push it into whatever I suggested as a candle and light it.

I also made a model to show how a candle works. The Newcastle chemistry department glassblowers made me a bunch of fine tubes in a little flask. I filled the flask with acetonitrile—an inflammable liquid that climbs well in glass tubes. In the studio, it was hard for the camera to see the flame at the top. In retrospect I should have added ethyl borate, to turn that flame green.

A Birthday Party to Remember

I once received a desperate cry for help from YTV. They were going to broadcast a birthday party for an 88-year-old. The take was planned for next day. Then someone realized that they could not afford the TV time to light 88 candles on the cake. Could I light the candles simultaneously? Foolishly, over the phone, I said yes. I told the producers that I'd have to design the cake and would provide a rig with candles on it. Later I heard that the company had some baker making a cake to my specification at 2:30 in the morning.

What do 88 birthday candles look like? I dashed into Newcastle and bought lots of them, and lots of the little flowery candle-holders they go in. I fear that many Newcastle tots had no candles on their cake that day, because I had bought them all!

My RIG had come up with only one idea to light the candles. It imagined a forest fire spreading through the inflammable leaf-bearing tops of trees. A chemical fuse (see chapter 9) might work, but I had to start at once and wouldn't have enough time to create anything complex. My instant design was a hollow square, made from four aluminum sheets by the chemistry department Mechanical Engineering Workshop at the University of Newcastle. Each sheet had a turned-up outer edge for rigidity, and the whole thing was bolted together to make one unit. To stop it looking metallic on TV, I painted it blue with spray-can paint. (To match that rig, I specified blue icing for the cake beneath.) Each strip held 22 candle-holders in two rows; I drilled the holes myself. They let me push much of each tapered candle-holder stalk through a hole; I stuck it in place by melting the protruding spike into a blob. I held all the blobs in place by spreading tape along the underside of the strip. It was all intuitive work. There was no time to be clever.

As I imagined things unfolding, the candles would be lit and then the whole rig could be lifted off the cake as one unit. But all the candles on the final cake had to be new. It would look odd and tatty on TV if some candles were old ones, with carbonized wicks, and others were new with white wicks. To experiment with some lighting method, I needed still more candles. I still had to invent a fuse for the candles.

I soon found that Jetex Igniter wick (see chapter 9) was far too slow. Sometimes it failed to light a candle; worse, it left a suspended metal wire between the candles, and this would show on TV. So I tried cotton sewing thread soaked in an inflammable solvent. This burned much faster. But the cotton often left an ash residue attached to one of the candles. If it swung down, it could act as a wick and light the whole length of the candle. I dashed into Newcastle again and got some other sewing threads. Polyester thread seemed hopeful. It left no ash residue, but if it melted it could stop the spreading flame altogether. A double thread, made by twisting cotton and polyester threads together, seemed to work. The melted polyester seemed to encapsulate the cotton ash. Double thread held more inflammable solvent, too. I put some water into each flowery

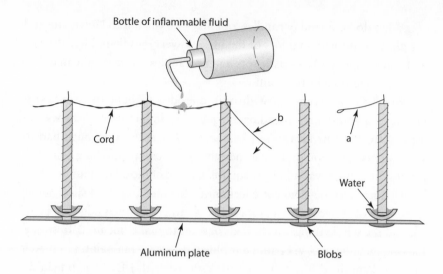

Bottle of inflammable fluid

Cord

b

a

Water

Aluminum plate

Blobs

FIGURE 10.2 Lighting Many Candles on a Birthday Cake at Once

The candles were arrayed on an aluminum plate, their candle-holders fastened in holes in the metal by melted blobs. The whole plate was then put on the cake. The wicks of the candles were all joined by sewing cord, which was wetted by inflammable liquid spread along it from a chemical wash bottle. When one candle was lit, the flame spread rapidly to all the others by burning along the cord. Cotton cord might leave an ashy cord behind. This could swing down and light the whole candle as at *b*; water in each holder put it out. Polyester sewing cord might melt as at *a* and fail to transmit the flame. A twisted pair of each type of cord worked well, and was better at holding the liquid from the bottle. I used this scheme twice, each time lighting about 100 candles simultaneously. It looked splendid on TV and saved all the bother and time of lighting the candles one by one.

candle-holder. Any fuse residue that did swing down would just go out, and the TV camera would notice nothing. My plan was to soak the fuse with transparent inflammable solvent just before the take, squirting it onto the fuse from a chemical wash bottle. Any drops that missed would simply form puddles on the blue-painted aluminum plate. The cameras would not notice, and the whole thing would be removed for the eating part of the show (fig. 10.2).

So I had something that seemed to work. I tried a lot of inflammable solvents on my thread—nothing toxic, of course. I finally settled on a mixture of toluene and petroleum ether (the former lasted a long time; the latter speeded up the flame. The two are easily mixed). My idea was that someone at the party would light my rig at a corner, then flame would spread rapidly around the hollow-square cake. And so I took this lash-up to the YTV studio. The nocturnal baker produced his cake; I put my rig on it; everything looked fine. I and an assistant kept putting my solvent mixture on the threads of the candles, using chemical wash bottles, right up to the moment of ignition. On the take all went well!

Later I offered to copy this technology for another party, the one hundredth birthday of Brian Shaw at Nottingham University. We were good friends. (I refer to his terrifying lecture on explosives in chapter 9.) As before, the event would be on TV. This time I bought even more candles and candle-holders. Furthermore, I had discovered that each candle lit more reliably if its wick was "fluffed out" with solvent beforehand, removing excess wax. I had also found a better solvent in which to soak my double thread: methyl ethyl ketone, which gave a speedy flame and lasted quite well. As before, I made an aluminum rig drilled with holes and painted it blue. As before, I melted the candle-holders in place. As before, I relayed to the cake baker details of the object I planned to put on his creation. But this time I put a Plexiglas fence around the rig. While transparent for the TV cameras, it let me put the candles out again, by pouring a bucket of carbon dioxide gas over them. In the event, my carbon dioxide was not needed. Brian Shaw, aged 100, blew the candles out!

Mirrors and Solar Reflectors

At ICI, I learned something about coating aluminum metal on polyester plastic sheets and then stretching it. The plastic extends much more than the metal; so the metal coating splits into a set of loosely connected microscopic islands. I was trying to make a sort of two-dimensional carbon-granule microphone. I failed.

Even so, I got a feeling for aluminum-coated polyester film. Polyester is a tough, important plastic. As a melt at about 250°C, it can be extruded through a thin wide slit into polyester film. To thin that film further, it is usually wound up faster than the extruder can deliver it. It is often stretched sideways during the extrusion process, or "tentered." This way of making film extends its thread-like polymer molecules in both directions and makes the film even tougher.

The resulting reel of strong, solid film is frequently "aluminized" in a separate process. It is put in a vacuum and unwound over a source of aluminum vapor, before being wound up again. This coats it with a thin shiny surface layer of the metal. Aluminized polyester film is made in vast amounts for the packaging of foodstuffs (that aluminum surface greatly impedes the inward progress of oxygen from the air, which can degrade some foodstuffs). I had a big roll of aluminized polyester film in my collection of junk. The roll was 1.2 meters wide but the film on it was only 0.012 millimeters across. It seemed ideal for optical play.

My RIG imagined a circle of it many centimeters across and further imagined reducing the air pressure behind that circle. The aluminized film would deform into a concave mirror. That small deformational stretch should not break up the metal coating. If the resulting mirror was an accurate paraboloid, you could use it as an astronomical telescope objective mirror, as used in the giant Palomar telescope. This notion was independently proposed by Peter Wadell and Bill King of Strathclyde University in 1985.[3] I never read anything more about it. I assume that they had the same sort of trouble that I did and gave up.

Before I even made a mirror from my film, I faced an intriguing technical challenge. Should the metallic coating be on the inside or the outside? I had met this problem before, while making a steam balloon, which needed its metal coating on the outside (see chapter 6). My imagined astronomical mirror also needed its metallic coating on the outside. It would then be "front-coated," like the aluminized glass mirror of the Palomar telescope. It would reflect all the starlight, infrared and ultraviolet as well. It would, of course, be very vulnerable to damage. I would have to be careful not to touch the metal coating at the front and not to let anything else touch it. And I planned to anchor my new mirror to a flat circular "rim" with an enclosure behind it. I could then apply whatever vacuum I needed to deform the film. The result would be a Palomar-type mirror with its metal layer at the front but polyester film at the back. It might be optically poor at the rim, but I hoped for optical perfection in the middle.

My first mirror used an old bowl, 52 centimeters across and with a smooth flat rim. The strongest adhesive I know is the very slow-setting form of Araldite epoxy resin. I mixed that glue with chloroform to make a paint, painted the rim of the bowl with it, and let the chloroform evaporate to leave a thin layer of viscous adhesive. Then I stuck my polyester sheet onto the rim. I waited two days for the epoxy glue to set. Then I tried applying a slight suction to the bowl, via a gas tap I had put in it. The epoxy glue held, and so did my mirror surface. Even slight suction deformed it! If I reduced the air pressure in the bowl by even a tiny amount, its flexible aluminized sealed film "lid" would deform from a plane mirror to an inwardly bent concave mirror.

That resulting concave mirror looked good to my eye (fig. 10.3). But detailed optical tests dismayed me. My converging mirror was a very bad paraboloid. The rim was indeed imperfect, as I had feared. But it was poor even in the middle. And a potential improvement suggested by my RIG—sucking the whole mirror violently to deform the edge region, failed totally. I could not even make the deformed film lie flat, let alone stick it down.

Worse, my air-sucked mirror would not hold focus. It gradually took up air, became less converging, and drifted over minutes into being a plane mirror. I looked for a leak; but I now suspect that the film itself lets air through slowly. It could not hold even a low vacuum for the long time needed by a useful telescope. I had made a vaguely converging mirror but not a telescope objective mirror that might compare with the costly Palomar masterpiece.

So, I could make a big converging mirror but not a telescopically useful one. What to do with it? I felt that I could concentrate sunlight. Further, solar power is overwhelmingly concentrated in the visible, where my thin polyester film was highly transparent. I would lose very little by putting its metal coating inside. It would then be "back-coated" like a commercial glass mirror. Unlike the Palomar mirror, a back-coated mirror is very robust. You can finger it but can never touch the metallic reflecting layer on the back.

My largest (and optically worst) mirror was 140 centimeters across. It could concentrate about 1.5 kilowatts of solar energy, and as a burning mirror, it could ignite a thick wooden plank in a few seconds! But big aluminized-film mirrors are not practical sources of solar power. They are very vulnerable to wind and rain; only on rare occasions (no rain, little wind, bright sun) can you use them in the United Kingdom

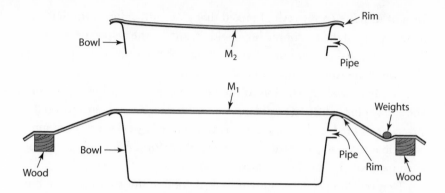

FIGURE 10.3 Aluminized Polyester Film, Stuck on the Rim of a Bowl and Sucked into a Concave Mirror

The lower picture shows a flat bowl about 50 centimeters across, its rim covered by epoxy glue to take aluminized polyester film 0.012 millimeters thick. An even thinner coating of aluminum on the top surface of the film, makes it a shiny mirror, M_1. A wooden surround, and several weights, hold the film flat and taut while the glue sets. A slight suck on a pipe, its gas tap not shown, then deforms it into a concave mirror.

Later designs (*top*) had the film with its coating underneath (M_2), to protect it. A slight suck on the pipe still made it a concave mirror. But I never found a region of optically paraboloidal surface. A non-paraboloidal reflector is useless for making a telescope, which was the original intent. But the big concave mirror was a splendid concentrator of sunlight.

at all. But I have exploited them in filmed TV demonstrations of solar power. I have even driven a little solar steam engine and shown how the sun makes rain.

So my cunning aluminized-film mirrors, while they failed to transform astronomy, still found some use. They may even find a place in my serious science. I have a thesis that, despite the claims of later historians, Archimedes could not have used solar mirrors to ignite a Roman fleet coming to attack Syracuse. Experiments are in progress!

Fun with Polarized Light

Polarized light can be fun, largely because it is not at all obvious to the eye. I provide a sketchy explanation of it in figure 10.4. Light, as we

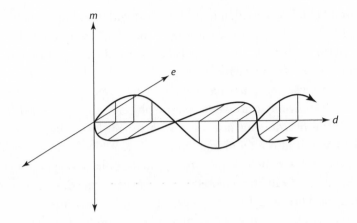

FIGURE 10.4   The Theory of Polarized Light
Light is electromagnetic radiation. A beam of it is a set of particles (photons)
each of which has an electric vector (e) and a magnetic one (m); the diagram
shows about 1.3 wavelengths of a photon d moving rightward. If every photon
in a beam has vector e in the same horizontal direction, the beam is said to be
"horizontally polarized." The eye cannot detect polarization. Most artificial
lights are not polarized. The vectors of their photons are evenly spread.

all know, is electromagnetic radiation. Any given "beam" of light contains
two vectors, electric and magnetic, each at right angles to the direction the
beam is traveling in. Sometimes a vector (the electric one, say), is particularly strong in one direction. Physicists say that such light is "polarized."
The eye cannot detect that polarization. To do so you need a special po-
larizer, such as the Polaroid film invented by Edwin Land. Much natural
light, like the blue light of the sky, is polarized; and so is light reflected
from a wet surface. Hence the value of "polaroid sunglasses" against the
glare from such surfaces. Conversely, light reflected from a dry rough
surface, and most artificial light, is generally unpolarized. The electric and
magnetic vectors of its beams are evenly distributed about their direction
of travel. Use a piece of Polaroid film to examine such light; no matter
how you twist the Polaroid, it will not change in brightness.

I liked the idea of making a strongly polarized lamp. It gave me a lot
of trouble; my first design (using a tungsten-filament bulb) got far too hot.
But I got there in the end. My final construction exploited Edwin Land's
brilliant invention of polarizing sheet. The lamp only emits horizontally

polarized light. You can illuminate an object with that lamp, and view it through a "crossed" Polaroid sheet (an "analyzer," admitting only vertically polarized light). Only light twisted by the object gets through. The rest of the scene remains dark.

My first use for polarized light was to demonstrate photoelastic stress analysis. Many transparent plastic objects—such as cassette holders, toothbrushes, drawing instruments—are made by forcing transparent plastic material into a mold and letting it set. While being molded, the plastic acquires a lot of locked-in stress, and the polarized light shows this up as an array of rainbow-colored bands. I made a rig with a plastic tape spool, a toothbrush, a T-square, a cassette lid, a protractor, a little container, a ruler, all made from transparent plastic—probably polystyrene or Plexiglas. Each shows the colors imposed by the stress of molding. If you heat a molded object, it goes back at least partly to the shape it had before being molded. (This is a form of the "plastic memory" effect, which I discuss in chapter 11.) In that relaxation, it loses much of its locked-in stress, as you can show with polarized light.

You can even stretch a flexible object in front of a polarized source and see the colored bands form and move! A polyethylene bag will develop bands that change interestingly while it is being stretched. But the polyethylene does not retract when the stretch is taken off, so they do not go back. A rubber condom, washed to remove the grease and thin enough to be transparent to light, will also develop colored bands when stretched in front of a polarized lamp. They even reverse when the condom is allowed to contract back; but the demonstration may infringe the bounds of good taste. Edible gelatin, essentially uncolored Jell-O, may make a better public display. But objects with artificial colors may interfere. Colored balloons and rubber gloves give a very poor show.

This photoelasticity is used seriously by engineers to judge the stresses on a structure or a component. You make a transparent plastic model of the component and subject it to the sort of load a real one would meet in service. You allow for the different elasticities of the plastic model and the engineering material.

Another use for polarized light is to look at crystals. Most solvents and solutions are entirely limpid and do not twist polarized light at all. But a crystal twists it strongly—indeed, early chemists used polarized

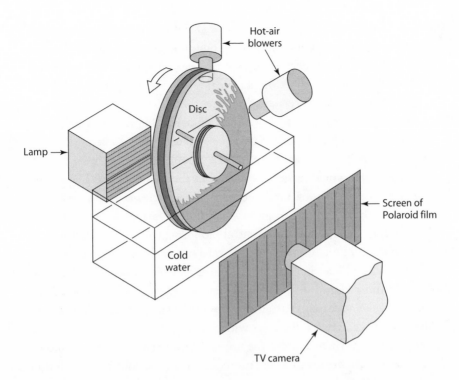

FIGURE 10.5   A Disc Using Polarized Light to Make Crystals Colorful for TV
The lamp emits horizontally polarized light. It goes through a glass disc, which rotates counterclockwise about once a minute and contains sodium sulfate dissolved
in water. Where the disc enters the cold water of the aquarium, its temperature
falls to 10°C, and it rapidly deposits sodium sulfate crystals, which, growing, twist
the plane of polarized light in a very colorful way. Where the disc emerges from
the aquarium water, hot-air blowers warm it to 30°C and the crystals dissolve
again. The colorful light twisted by the crystals gets through the vertical polarizing
screen (made of Edwin Land's "Polaroid" film) and the TV camera sees them. This
display can run for many minutes, showing the continuous colorful crystallization
of sodium sulfate, which to the eye is a boring white powder.

light a lot to spot and identify crystals. I played around with this effect a
great deal. I wanted to show colored, polarized crystals on TV, using my
polarized lamp. The compound I finally settled on was sodium sulfate.
This colorless crystalline salt is easily available in quantity. It has two
properties I liked. First, it is massively soluble in hot water and much less

soluble in cold water. So I could grow or shrink its crystals just by changing the temperature of its solution. Second, it does not twist the plane of polarized light very much. My apparatus (fig. 10.5) displayed a solution about 0.5 millimeters thick. In chemical terms this is a great deal; but I could not make my TV "crystal disc" any thinner.

Even when my RIG had dreamed up a feasible design, I had a lot of trouble making it. I wanted two glass plates about 40 centimeters in diameter and 0.5 millimeters apart. The 1 millimeter sheet used for microscope slides would have been ideal; but the thinnest glass I could get commercially was about 3 millimeters thick. My glazier could cut it into discs, though he declined to put a small hole at the center of each disc. Accordingly, I could not put a central mounting rod through both discs. I had to mount them on two coaxial rods, one each side: a much weaker arrangement. I liked the idea of sealing the two discs together with some sort of rubber clamp seal, but in the end I had to use silicone resin (as used to make aquariums) around the outer edge. I left a small hole and filled the space between the discs with hot sodium sulfate solution, using a hypodermic syringe. I sealed the hole with more silicone resin. A motor rig, with a variable speed drive, turned the disc at about one revolution a minute. I played with that rate of rotation, and the heating and cooling of the disc, so as to get the best display. It was easy enough to cool the turning disc: an aquarium tank holding cold water would do it. Warming the part in the air was more trouble.

Right up until the last moments I had the idea that the disc should turn clockwise and that the main display would be crystals dissolving as the solution warmed up. But when I started to play with the rig, I realized that counterclockwise rotation made it far prettier. The left section of the solution then suddenly entered the cold water I had provided to cool the turning wheel. The solution inside it cooled fast: it deposited crystals that grew rapidly, gorgeously and colorfully, and dissolved again as that section of the wheel emerged from the cold water and began to warm up.

I wanted to cycle my disc between about 30°C and about 10°C. The second part was easy—I merely had to control the temperature of the water in my aquarium tank. The first part was much harder. To warm the right-hand side of the disc as it emerged from the water, I had to use two hot-air blowers. I disliked their noise, but no silent heater worked well. The whole disc gave a lovely continuous crystal display and worked so beautifully that I had in mind to turn it into a continuous crystal-chemistry exhibit for a science museum. But sadly, silicone resin is permeable, not merely to water but to air. Bubbles got into my disc and it slowly deteriorated.

11

Properties of Materials

Most of us just take the properties of materials for granted. Glass is fragile, while plastics are flexible. And yet every substance somehow reflects the structure of its molecules. As a chemist, I have worked all my life with glass; at ICI I learned a lot about plastics. I and the company tried to do a lot with those products, mostly foolish extensions or changes to their "natural" properties. Yet in some way those properties and the reasons for them got "downstairs" to my unconscious mind. Later on, I found myself appreciating the nature of materials and wanting to show some of it to a wider audience.

Bending and Breaking Glass

I have broken a lot of glass, mainly by accident. And yet I admire the stuff greatly. Bent too far, it breaks, but, bent slightly, it is amazing elastic. And yet to the public, glass is just that "fragile stuff."

My experiences with glass may have stimulated my RIG to play with its properties. Thus we all know that a ball or a stone thrown through a glass window makes a spectacular shatter pattern. Its straight cracks radiate out. By contrast, a bullet fired through a window (which many of us have seen on film) makes a small hole. I assume the bullet is traveling at a speed approaching that of sound in glass and is through before cracks can accelerate. I once tried the experiment. When I was a young lad, the Jones family dismantled its old TV set. Its cabinet made a house for Creature Jones; Mum made a cushion to make it more welcoming for him. The cat loved it (you can see him in it in the frontispiece), and I looked over the electronics for parts I could use.

The cathode ray tube was too big and clumsy for any project I had in mind. I decided to fire a steel bearing ball through its face, using a gunpowder-powered cannon I had made—a highly illegal object. The big old cathode ray TV tubes had a vacuum inside, so that the internal electron beam could move freely; this big volume of vacuum worried me a good deal. I put the TV tube in the family trash can, which had been set on its side for the experiment. In retrospect I should have emptied that can and been more scientific. But I was lucky; my lash-up worked. I merely propped the neck of the TV tube on an empty can that I found in the trash and covered its face with a piece of fabric sacking before lighting the fuse of the cannon and running away. The bang was probably mainly the cannon going off, but the impact, destruction, and sudden implosion of vacuum must have added to the noise.

The result puzzled me. The tube was largely intact. Its neck had come off, probably because it had been thrown back against the rear of the trash can. It had a hole in its front face, where the bearing ball had entered. Damage had spread a few centimeters from that entry hole; perhaps the ball had not been going fast enough to give a really smooth one. And there was not one, but *two* exit holes in the back of the tube! I scrabbled in the bin to find the exit objects, and deduced what had happened. The piece of glass punched out by the ball going in had also come out!

This risky experiment sharpened my interest in the breakage of glass. Automotive glass is toughened against breaking. While the sheet is hot and soft, cold air is blown on it, so that the surface cools and hardens first. Indeed, the surface is already solid when the bulk cools. That interior contracts, as almost all substances do as they cool. It puts the solid surface into compression and itself goes into tension. Glass is stronger in compression than tension, so the result is a sheet of glass with a hard compressed skin, difficult to break. If the toughened glass windshield of a car is broken, say by a stone thrown up by the vehicle in front, the whole sheet shatters. Every part of the glass sheet has that locked-in stress and spreads the damage on to the next part. The windshield shatters into many fragments and goes opaque but does not fall apart.

This once happened to me. Suddenly my whole car windshield shattered and went opaque. Fortunately I knew what to do. You punch a hole in the shattered windshield and can then see out of the hole, well enough to halt the car, anyway. The fragments of windshield, each a couple of millimeters across, do little damage to your fist. They are not exactly rounded, but they do not have the cutting edges of normal broken glass. Indeed, criminals often break a window to get into a car or vandalize it. I have often seen a pile of glass fragments on the pavement, where some vandal has broken a car window.

And yet, within its range of deformation, glass is highly elastic. I first met that elasticity as an undergraduate. I had a float supported on a glass spring and measured the density of various gas mixtures by the degree to which they buoyed up the float. The experience stays in my mind. Much later I met the elasticity of glass again, while screwing mirrors on to a steel frame. Each mirror, a couple of millimeters of glass with a reflective coat, bent! It distorted whatever it reflected. When I unscrewed a mirror and relaxed its stress, the reflection came right again.

I even expounded the elasticity of glass on TV once. I had the idea years earlier, when I bounced a glass marble on the family fireplace. For a Yorkshire TV episode, I bounced a marble on a ceramic tile. To make the item work, the tile had to be strongly cemented to a brick beneath it. The producer and the audience, most of whom thought of glass and ceramic tiles as inherently fragile, were greatly impressed.

Glass is not actually a solid. It is an extremely stiff liquid. It was discovered by accident, has been known for thousands of years, and has made possible several important technologies. Thus chemical apparatus is almost all glass; optics needs glass for lenses in cameras, telescopes, microscopes, and eyeglasses. Sheet glass allows that triumph of modern architecture, the glass window (I muse on it further in chapter 15). And glass can be amazingly elastic. In the nineteenth century, Sir Charles Boys melted a rod of silica, a kind of glass, till it was floppy. He held one end and fired the other from a bow, like an arrow. The rod stretched right out and the resulting fiber was sometimes too thin to find. But it was highly elastic; when found, it made a uniquely sensitive torsion balance. Modern optical glass fibers are still made by stretching a heat-softened glass rod; but these days you don't fire them from a bow. And you don't use them in any balance; instead, you shine light pulses down them. A typical optical fiber is only about 0.1 millimeters across but replaces many copper telephone wires. It is only tractable because it is elastic. It can be coiled up on big drums and uncoiled again at the site of use.

I have since mused on further demonstrations of the properties of glass. For example, how fast does a car window crack? I do not know, and I have not asked a vandal. I once shattered a car windshield over a meter long by laying it on the ground and hitting one end. It all seemed to break simultaneously into granules a few millimeters long. So the visual presentation on television might need to be slowed down a lot.

Memories Are Made of Plastic

How to get an audience interested in plastics? To most people they are just boring. But try heating them! The thermoset resins (such as Bakelite) are like wood or paper; they don't melt but char. But many thermoplastics (such as PVC, polyethylene, polyester, polystyrene, and nylon) undergo a "glass transition." They go from a "glass" with a well-defined shape to a "rubber." Even neater, that rubber "remembers" the shape from which it was stretched.

Now many commercial objects are molded as rubbers and remember their original shape. Thus the tubs used to contain foodstuffs are often "vacuum formed" from sheets of plastic. You heat the sheet up to soften it, "vacuum form" it by sucking it into a mold of the tub, and then cool it, when it holds that shape. Later you may print on it.

I knew much of this technology and chemistry vaguely, when I began searching for plastic objects to use in TV or lecture demonstrations. Heating commercial products is always good theater, but I was looking for something that (a) had some scientific message and (b) would be familiar to a lay audience. I bought and heated bank cards, yogurt containers, telephones, records, and many other plastic consumer products. Gradually I learned the practicalities and developed some demonstrations.

For me, food tubs have a great appeal. My ideal tub had been formed from a sheet. When heated again, it would not just go floppy; it would remember the disc shape it had before it was vacuum-formed and would contract back to it. Such plastic objects made a splendid lecture or TV demonstration. I bought lots of some products, purely for the tubs they came in. In one demonstration, I heated a tub with a butane camping-gas burner, or (better, because more controllable) a hot-air paint-stripping blower. Properly heated, even a deep tub went back to the disc from which it had been formed. Any print or picture on the tub was deformed with it. The result was often delightfully weird! The TV camera could linger on it closely. One of my greatest finds was a Danish margarine, of which both the tub and the lid deformed to a disc in boiling water. I bought a lot of that margarine.

The glass-to-rubber transition is well shown by PVC, as used in bank or credit cards. These sacred objects are routinely cold-stamped with names and numbers. It is splendid theater to take such a card out of your wallet and drop it into boiling water! This takes the PVC into its rubbery state. It loses all its embossing and becomes quite floppy. Indeed, I built a special gadget to heat up such a card and stretch it. Sadly, the trick is not entirely reliable. At certain digits (such as closed ones like 0) the card may puncture and tear from weakness. But if it works, it is a fine demonstration. You can remove the elongated card from the stretcher and show the audience. Even better, the stretched plastic remembers its original shape. Heat it again, and it will contract back.

The reversible glass-to-rubber transition in PVC gave me another idea too. I used plasticized PVC flexible tube. In the cold this stuff is stiff, though it can be pushed onto glass or metal piping. But if you pass steam through the tube, it heats up and becomes a true "elastic." So my cunning catapult used a steam-heated PVC tube. I planned to use this as a scientific display for Westdeutscher Rundfunk (WDR) TV. I planned to project a glass marble from it. But how to make this suitable for TV? No TV camera could see a small fast-moving marble. I had planned to fire my marble through a paper target, when the camera could see the sudden hole. But quite by accident, the German studio had a band in it. In a rehearsal I hit a drum with my marble. It gave out a deep "boom!" that greatly improved the television demonstration. It was a real aha! moment. The drum became a crucial part of my demonstration!

I had a lot of trouble aiming and firing my marble-firing catapult. Much of it was too hot to hold, yet I had to haul back the cradle that held the marble. And we had to anchor the steam kettle very firmly, against my involuntary starts and spasms. But I could see no way of avoiding that kettle. Steam is obviously hot, it was obviously going through the PVC pipe, and it made ideal TV. I never seriously attempted to make an

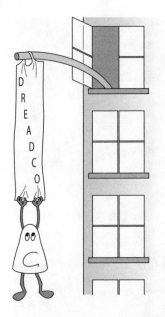

FIGURE 11.1   A Sleeping Dream Recalled by the Author
In the dream I had to jump from a window. I limited my rate of descent by stretching a plastic bag. A disembodied voice intoned, "Commend your soul to Woolworths."

electrically heated catapult with no visual impact. To avoid scalds, I wore gloves. And that drum did give it splendid drama!

Another plastic demonstration came to me in a dream (indeed, it is the only dream I have had whose technical content survived significantly into the real world). I was in a tall building, which may have been on fire. Anyway, I had to escape from it. I got a plastic bag and anchored it to something. I then hung it out the window and threw myself out clinging to it. A disembodied voice said, "Commend your soul to Woolworths." The bag stretched and let me down safely (fig. 11.1).

In the waking world, I took that dream seriously. I started to play with bags and plastic sheets, many of which can stretch to five or six times their original length. Sometimes they even get stronger in the process— their spaghetti-like molecules are aligned by that stretch.

The plastic that is made into bags is usually made by a wonderful process called "blowing," in which gooey melted polyethylene is squirted through an annular orifice and inflated. It expands into a continuous film of "lay flat tubing," which is flattened and rolled up as it forms. Its thickness is determined by the blower set-up. The polyethylene tubing for my descender was some 0.13 millimeters thick. I made it from commercial bags. I wanted to stretch my bags sideways, as lay flat tubing is more

FIGURE 11.2 A Plastic-Strip Descender Inspired by My Dream
A feasible descender needed many plastic strips, so that if one broke the others would do the job. I cut them from a commercial plastic bag. In the photograph, my nephew, Tommy, is descending, while my niece Philippa is watching enthusiastically. Having proved the descender, I used it on German TV.

stretchable that way. But a single bag was unsafe: it might tear across completely. So my "descender" used many polyethylene strips, each a few millimeters across. I cut them sideways from my bags, getting circular strips about a meter around.

Each strip was a simple plastic loop, with no end. I threaded the proper number of strips between two wooden rolling pins, each with a steel axle. The upper one was anchored at the top; the other would be suspended at the bottom, and the intrepid human user would cling to it. The number of plastic strips had to be just right. Too many for the user's weight, and they would not stretch at all; too few, and they would break.

Just right, and the human user would come neatly down. He would stop cleanly at the bottom as the stretched loops gained strength and declined to stretch further.

I asked my nephew and nieces to test the thing out (fig. 11.2). They were delighted to try the crazy falling rig their uncle had invented. My brother and sister-in-law were less delighted but went along with the idea. Later, I used it on German TV. Alexander von Cube of WDR Cologne posed a question about plastic molecules and said to his interlocutor, "No master will fall from the sky"—a widely known German saying. Then I descended from a special tall rig on my plastic strip device!

Boiling Water in a Plastic Bag

I have lots of thin plastic films and laminates. My aluminized polyester film inspired the solar concentrator (see chapter 10) and the steam balloon (see chapter 6). When anything with film in it worked, I often got a lot of that film later, to do it again. But much is scrap and junk.

I am in favor of junk. Over the years it may have inspired my RIG. Thus I once heard of a scheme to boil water in a paper bag. The water stopped the paper charring, and the flame evaporated any water leaking through the paper. I thought it would make splendid TV. I wanted to do it with a plastic film. First, like paper, plastic film is not an obvious boiler material; second, you can see the flame through it. So I started to experiment. I scalded myself with liters of hot water and destroyed lots of plastic film.

I soon discovered the snags. For a start, rigid films were no good. You had to fold them to make them hold water at all. And in my experiments, the outside of a fold is heated by lots of flame, while the inside is inadequately cooled. The film splits along the fold. So I began to play with deformable films such as polyethylene, polypropylene, rubber, or nylon. Rubber film, of course, deforms by the mere weight of water in it (I used a rubber balloon and a condom). And I could deform a pliable film, such as polythene from a bag, by forcing it over an old rounded wooden hat block that had belonged to my mother.

I still failed, but I studied the failures carefully. Each film went into little holes, which leaked. And I reckoned each hole had been the site of an adhering bubble—which of course conducts heat badly. The simple answer was to use boiling water; boiling gets rid of dissolved air and bubbles. It makes for better theater too. Most of my films were still very likely to spring a leak. But some nylon Alcan roasting bag film that I had was easily deformable on my hat block, and it seemed unaffected by boiling water or a flame beneath that water. So I built an apparatus in which deformed nylon film could be clamped with foldback paper clips into a ring made from an old metal magnetic-tape spool. I could pour boiling water into the film and then heat it with a butane camping-gas burner. It boiled the water convincingly!

I got away with it on German TV; but the demonstration later failed in the most embarrassing circumstances. I was showing it in a Royal Institution lecture. At the time I feared that the flame was too hot. Later I used a cooler flame: methylated spirit in a little burner. This worked, but then so did the hotter butane burner. I still do not know what went wrong.

12

Physical Phenomena I Have Noticed

I have spent a lot of time just playing with physical effects. I recommend this random play. It may be part of curiosity, which I salute in chapter 1 or it may help to develop physical intuition (see chapter 4). Thus, for no good scientific or technical reason, I have been distracted by the noises of steam and by the way certain objects can be levitated on an airstream. I have also played with the "convective cells" that form in a heated viscous fluid and with optical phosphorescence. At the time, I understood all these things badly, if at all. Yet I contrived to turn them all into demonstrations worth showing to an audience. I even learned in the process!

The Noises of Steam

Early in my career as an amateur steam technologist, I began to be aware of the noises that steam makes. I tried blowing it into water. It made a rapid banging, a lovely noise. Even better, the noise changed its tone and character as the water got hot. Ultimately the water reached its boiling point, and the noise changed character quite suddenly.

I understood (maybe for the first time) something I had learned about in my physics lessons. Water exerts a vapor pressure, and the hotter it is, the greater is that pressure. But below the boiling point, it is always less than atmospheric pressure. So, cooled in water, a steam bubble condenses very sharply: it makes a bang as it contracts. But it also gives out latent heat and warms the water. The more steam you blow into water, the hotter the water gets. The bubbles of steam condense more slowly, and their noise changes. When the water reaches boiling point, its vapor pressure equals atmospheric pressure, and the steam no longer condenses. It just goes through the liquid like any other gas, making only a vague bubbling sound. So at the boiling point, the sound of steam changes abruptly.

Now I suspect that many of these experiences went "downstairs" for the RIG to play with. I gained a sense of steam as a physical medium. They probably meshed with my foolish but interesting scheme to make a steam balloon (see chapter 6). Years later, thinking about steam in television terms, I realized that nobody looks at an electric kettle. They listen to it.

All this began to form a possible TV demonstration in my mind. I started to listen carefully to electric kettles. I decided that there were three noises to invite a TV audience to listen for. When you fill an electric kettle with cold water and turn it on, it gives an initial quiet fizz as air is expelled. Later there is a characteristic crackly, banging noise that changes as the water warms up. When the water boils, the noise goes suddenly much quieter and changes its character too. I could show these effects quite well with my steam-blowing system, without simply copying a kettle, which nearly every viewer would already have.

I set up the kettle scheme for Yorkshire Television Ltd.; it made a neat item for Magnus Pyke. Later, I developed the idea further for Westdeutscher Rundfunk Cologne. I even managed to tell a human story— which I got from an employee at the Science Museum library. That library was heated by cast-iron steam radiators. When the steam was turned on, they warmed up with a fearsome banging noise. Their previous service had put water in them; this had cooled but was then heated and expelled by the fresh steam. In the 1930s, many refugees from European dictatorships came to London. Some of them worked in the Science Museum library. When the heating came on, and the radiators began their terrible banging, many refugees thought it was gunfire. Violence had come to London! Some of them hid under the tables in alarm. I even managed to illustrate my story with real sound! Those old steam radiators were still in use in the 1960s, and I made a recording of them while renovating an old portable BBC magnetic-tape recorder.

Levitation

A light beach ball can be levitated on the updraft of air from a vacuum cleaner used as a blower. The experiment is delightfully counterintuitive. You would expect a ball, or indeed any light object, to be blown away by an updraft. Yet there it stays! I was fascinated and later tried it myself. I made a light beach ball levitate on the updraft of a vacuum cleaner driven as a blower, varying its driving voltage with a controllable transformer. I found that levitation would work over a wide range of blowing efflux. A tapering nozzle on the pipe outlet seemed to help things along by speeding up the local gas flow.

Beach balls from 16 to 27 centimeters in diameter all seemed to work. Adam Hart-Davis, then of YTV, tried a light, circular aluminum Jell-O mold 21 centimeters across and that worked too. We used it in a YTV film of a "flying saucer"—the vacuum cleaner and piping being out of shot, of course. Later I experimented with all sorts of light non-spherical objects, to see what would fly. Even a bun flew, but my favorite was a plastic bottle (fig. 12.1).

What is going on? The levitating ball is an example of the Bernoulli effect: gas flowing in one direction exerts less pressure in the other directions. A draft in one direction, such as the efflux from an air blower, implies a reduced pressure in the other directions, and this stabilizes what the efflux is lifting. How strong is the force that holds a levitating object in place? I guessed 10 grams, but an experiment found up to 70 grams. My physical intuition had been too pessimistic.

When I showed the effect to a German TV audience, I used a non-spherical floater: an inverted Frisbee. I balanced it a vertical pipe connected to a blower. Then I turned the blower on and slowly built up its voltage. The Frisbee rose in the air and levitated stably in the updraft from the pipe. Then I slowly turned the blower voltage back to zero. The Frisbee sank back and sat stably on the pipe. The studio audience was entranced!

When something works well, my normal instinct is to scale it up. But scaled-up gas demonstrations often go turbulent. So I scaled my Bernoulli system down. I tried to lift a ping-pong ball on a hair drier with a disconnected heater. This worked well; even more amazing, I could levitate two ping-pong balls, one above the other! I guessed that the updraft closed around the first ball and reformed into a jet that could lift the second. My RIG had the immediate fantasy of a whole string of vertical balls suspended on one jet.

I never got this to work. I even failed to levitate small, light, spherical Styrofoam beads, also known as "peanuts," on a jet a few millimeters

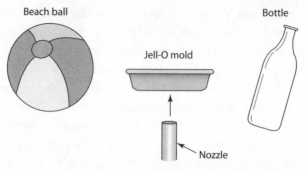

FIGURE 12.1   An Upward Jet of Air Can Lift Things by the Bernoulli Effect
An updraft (as from a vacuum cleaner used as a blower) can levitate a spherical beach ball. But it can also levitate an aluminum Jell-O mold—making a "flying saucer"—or a light plastic bottle.

across. My fantasy of a lecture or TV demonstration, in which a tower of balls or beads is levitated on one gas updraft, remains just that, a fantasy.

Convective Cells

A heated liquid expands and rises and the cooled liquid contracts and falls. The movement of the hot in relation to the cold may result in a convective flow of the whole liquid volume or, with more viscous liquids, ordered "convection cells." Aluminum powder can make the cells visible. It can come as microscopically small flat metal plates, and convection physicists often add it to their fluid. I tried it. I failed to predict what cell pattern would form, but one always did. Anybody who heats viscous liquids—in effect, every cook—would be interested. I started to imagine a TV demonstration based on the idea and began to play with viscous liquids. Olive oil was not really viscous enough. Glycerol is available in gallons but would just mystify a TV audience. So I went for one of my favorite liquids, golden syrup.

Now you can't just mix syrup and aluminum powder. The powder sits on top of the syrup and refuses to mix in. So I slurried my powder in a little water with a dash of detergent. Then I added golden syrup, ultimately in vast excess. When I had stirred this mixture and had left it for the bubbles to rise out, I began to play. Even in a heated frying

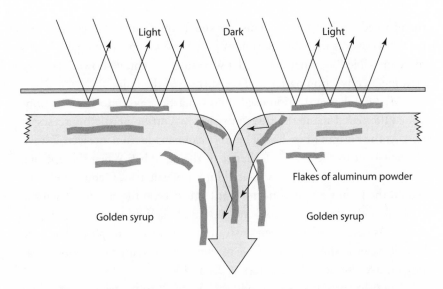

FIGURE 12.2 Convective Cells in Heated Syrup Made Visible by Adding Aluminum Powder

As a heated liquid expands, the hot liquid rises and starts a flow called "convection." If the liquid is viscous (like golden syrup), it forms a set of invisible "convective cells." Convective flow occurs within it. The junction between two convective cells can be made visible by tiny aluminum plates suspended in that syrup and shown in the drawing as long thin rectangles. The particles of aluminum powder are often tiny flat sheets. At the left and the right, syrup flows horizontally along the surface of each convective cell, toward the middle where the two cells meet. It carries the tiny aluminum plates horizontally along with it. Incident light reflected by those plates (shown by the word "light") makes the surface bright. Where the two cells meet, the syrup goes downward (shown by the arrow). The tiny aluminum plates turn along with it, and (as the word "dark" indicates) the junction looks dark. A viewer looking down on the liquid surface sees the convective cells as a pattern shown up by their dark junctions.

pan, it created very obvious bloblike convection cells. It had great TV potential!

I planned to heat my mixture electrically. My electrodes were bent stove elements. I had a problem: how do you bend a stove element without it snapping? Since copper alloys, if heated and suddenly cooled, often go soft and flexible, where I wanted to bend an element, I heated it and then cooled it under the tap. This worked. I could put an "electrode" in a tray and cover it with my aluminum-powdered syrup. Submerged, it was quite invisible. But when I passed a warming current, it showed itself by creating a convection pattern on the surface of the fluid.

My final demonstration used yellow plastic trays, into which a camera could peer. I made the trays in pairs, by cutting plastic briefcases in two. At the bottom of each tray I put a heater, made from an electric stove element bent to the shape of a letter of the alphabet. When the opaque, aluminized syrup liquid was poured into a plastic tray, it completely obscured the heater at the bottom—until I turned on the current. Then the concealed element warmed up and revealed itself by its convective pattern. This made a splendid demonstration, and WDR was pleased to have it on its science show. In deference to that company, my three demonstration electrodes formed the letters W, D, and R.

My other problem was getting the materials to the television station. I made up about 5 liters of my golden syrup and aluminum powder mixture and put it in an empty plastic oil can. I put all my gear in a cardboard box and gave it to the air-freight people, for shipment by air to Germany. The customs bureaucrats must have been used to the strange consignments I sent to Cologne.

Then I remembered a story my Dad had told me about an advertising campaign for shampoo. At the last minute somebody reckoned that more trial-size shampoo sachets were needed. So a large crate of sachets was rushed out to the campaign region by air freight. Now a cargo aircraft is not pressurized. So as the plane climbed toward operating altitude, the air pressure inside it fell and the shampoo sachets expanded. Many of them burst. The released lather continued to expand ominously and emerged through seams in the crate. One of the aircrew glanced back and saw a mass of lather advancing on them from the cargo space. The plane radioed an emergency, and put down hastily at the nearest airport.

Now I had no idea whether my air-freight consignment was going to Cologne in a pressurized plane. I phoned the air-freight people, but they could not tell me. They had sent my consignment to Manchester Airport by road; the flights from there to Cologne were beyond their knowledge. But my container had a lot of air sealed in it. It was like a huge sachet. I had terrible visions of my syrup and aluminum powder mixture bursting out in some awful explosion. I did not think it could do any damage, though my fearful imagination painted ghastly images of Lufthansa 659 falling out of the sky with syrup in the hydraulics. So I got another oil can of the same type and fitted its cap with a pressure gauge. I set the whole thing up out of doors and put water and solid carbon dioxide in it (this mixture would build up the pressure). I screwed the cap on. Through binoculars I studied the rising pressure gauge on that oil can and took a telephoto photograph as it blew up. It failed at 1.5 atmospheres by the shearing of the filling cap screw. Now 1.5 atmospheres is well beyond the internal pressure that could be imposed on the cargo of an unpressurized aircraft. I concluded that my can would easily survive the rigors of its journey! Even so, when I got to Cologne a few days later, I was very pleased to find that my crate had arrived safely. My demonstration gear was not covered in sticky metallic syrup; neither the transit case nor the syrup container itself was ruptured. And the convection demonstration worked!

Paints and Phosphors

Paint is a splendid invention. It can be stable for centuries: hence such revered masterpieces as the Sistine Chapel ceiling. Conversely, fugitive paints that slowly lose or change their color over time are big trouble for artists. Daedalus began to imagine a paint that changed color when the painted component beneath it was stretched or deformed.

All engineering depends on the rigidity of solids. When a structure carries a load, every part of it stretches or shrinks a bit. In a fully laden structure, the most stressed part may change size by as much as 1%. So Daedalus began to muse on a paint for load-bearing structures. Even small size-changes would make it change color. Now paint is pigment particles suspended in a liquid vehicle. Once it is set, it has to stretch or shrink with the component it has been painted onto. There seem two options—either the pigment particles stretch or the set binder stretches. Daedalus mused about the first option first.

His 1976 invention of Stresspaint pointed out that many chemical reactions need a highly specific catalyst.[1] Ammonia synthesis, for example, uses an iron catalyst, carefully doped with the metallic element molybdenum to get the atomic spacing right. Daedalus felt that the tiny stretch of a paint particle might make it catalytic. So he set DREADCO's chemists to inventing a color-changing particle reaction.

This catalytic change was permanent. Once triggered, it could not be reversed. Aircraft structures, which have to be replaced if they creep beyond some specific limit, would greatly benefit from Stresspaint. That permanent change of color would show that at some time the painted component had been loaded to the extent revealed by the changed color.

Later Daedalus imagined a paint whose particles (if there were any) did not change. Instead, the binder stretched. Paints with no suspended particles, but with a dyed binder, are called lacquers. So the new product, unveiled to the public in 1988, was called DREADCO's Loadlacquer.[2]

To create Loadlacquer, DREADCO's chemists were told to make dyes whose molecules were very crowded. Color depends sensitively on molecular overlap, so even a slightly stressed Loadlacquer should change color dramatically. When the load came off, the lacquer would return to its basic color again. When you walk over a bridge, it sags slightly beneath your weight and springs back when you have gone. So a Loadlacquered bridge would change color locally as you walked over it and increased the load it was carrying. When you had gone, it would return to its original color. Any excessive loading would signal its own warning; bridge engineers would rejoice!

Much later, I realized that I had made the perfect material for Stresspaint and Loadlacquer while still a chemistry student. Indeed, it may have been in my unconscious mind all the time. I still recall my amazement on making and playing with Fremy's salt, $(KSO_3)_2NO$. It is a yellow solid that forms a blue solution. The molecules in a solution are more widely spaced than they are in a solid. I love the idea of turning Fremy's salt into some sort of Loadlacquer in which it could traverse its chromatic range more gently and reversibly.

A related Daedalian scheme also broke new optical ground. Many solar-energy schemes fail at night. The sun goes down, just when we need its light! So Daedalus began to think about storing daylight and releasing it at night (see chapter 15). Many substances, like calcium sulfide and zinc sulfide, can store light. Optical energy forms stable electron-hole pairs in the crystal. Later they combine and emit light again.

Many kilograms of plaster are needed to coat the walls of a room, so in 1973 Daedalus proposed a phosphorescent plaster.[3] He called it Phaster. Ordinary plaster is calcium hydroxide, and plaster of Paris is calcium sulfate. So Phaster might be, at least as a pilot recipe, a mixture of calcium hydroxide and calcium sulfide. The human eye can work on a tiny fraction of full sunlight, so the principle is not absurd—at least not for Daedalus.

These cunning "optical illusions" did not come purely from my RIG. They were stimulated by experience—you can actually get phosphors, such as glow-in-the-dark stickers. I managed to buy some calcium sulfide phosphor powder and some "phosphor sheet" and played with them. Exposed to light, phosphors absorb a bit and later glow greenish for a few seconds. They cannot store anything like the amount of light that Phaster would need, nor can they store a wide range of colors; but my play suggested a demonstration for German TV. The largest phosphor sheet I could find was 44 centimeters by 35 centimeters. I put it in a neat frame. Then I made an owl in copper pipe, a tricky piece of soldering. (The German WDR science show had an owl as its symbol.) I fastened that owl on the back of the phosphor sheet.

In the TV studio I poured boiling water into the piping, when the copper owl got very hot. So did those bits of phosphor sheet in front of it. When I illuminated the front of the blank picture with a big photographic flash-lamp, it absorbed a bit of the sudden bright light. The hot regions of the phosphor disengaged its light very fast. The owl stood out in bright positive green light. It was soon depleted of optical energy compared with the cooler phosphor around it and went negative as you watched. It made very pretty TV!

I also played with my calcium sulfide phosphor powder more personally. I wanted to show my nieces and nephew something about the technology of cleaning. I asked my long-suffering brother and his wife to let me play with their sofa in the dark. I sprinkled it with calcium sulfide phosphor powder. When I flashed it with a photographic flash-lamp, my nieces and nephew saw that the particles of phosphor lit up like glowing stars all over the sofa. I asked them to clean it off. Of course they attacked the sofa vigorously with dusters and brushes; but they merely spread the phosphor grains around. When I flashed the sofa again, most of the glowing grains were still there. Conventional "cleaning" had not removed the dirt but had merely smeared it about.

Then I attacked the sofa with a vacuum cleaner and sucked most of the phosphor grains away. Despite all my efforts, some remained behind. My brother and his wife still have a feebly phosphorescent sofa.

13

Odd Notions I Have Played With

My creative life has constantly had me chasing ideas to see where they go. Often I have developed some notion without a feasible commercial market or indeed any outlet at all. Should I just make a note and pass on? The scheme might come in handy later. But sometimes an idea is too appealing just to leave. Thus when I thought of my hypochondria notebook, I developed it and ran it in parallel to the medical survey imposed on me by the UK National Health Service, or NHS. Tetrachromatics started life as part of a scheme for television; it failed, but I continued to like the idea. And the basic notion of non-chemical chemistry—randomers—gradually grew up over my life as a professional chemist. I ultimately used it in the Daedalus column but continue to feel that it may not be total nonsense. So it is part of this book!

The Hypochondria Notebook

I started my hypochondria notebook in 1975. It was based on two ideas. First, it was to be my own medical record. I would note the form and time of onset of all medical details: aches, pains, and general symptoms. Most would be trivial and would just go away. But if anything grew into a serious medical condition, I could look back and see how the thing had started. Second, in the notebook, I treated myself as a biological specimen, one about which I had the most intimate knowledge. Among the things I noted were my body temperature (taken with a medical thermometer, under my tongue as Mum taught me). It varied slightly throughout the day, so I took it at different times of day and made a graph. I could thus tell when it was out of line with expectation, even slightly. Day by day, I noted my weight. To minimize scatter, I always took this under the same conditions: naked, in the morning, after urinating, before breakfast.

In 1988 I got a blood-pressure monitor and often recorded its findings. As a good scientist should, I regarded it with great skepticism. I calibrated it against absolute pressure standards and compared it with a different sort of monitor. Any blood-pressure monitor uses the same principle. You wrap a cuff around your upper arm, at about the same level as your heart. The machine pumps the cuff up to a high pressure with air and cuts off the blood flow in that arm. Then it gradually reduces the air pressure in the cuff. Both of the monitors record the pressure at which blood first starts to flow in the arm—the maximum systolic pressure of your circulation. One type of machine uses a microphone; the other senses expansion pulses in the cuff. As the applied pressure declines, each machine notes the lower, diastolic pressure at which blood keeps flowing in the arm even against the applied cuff pressure. A typical value might be 130 mm Hg systolic and 85 mm Hg diastolic (mm Hg = millimeters of mercury, from the days when a small mercury manometer was used to read the pressure). The machine also records the pulse rate, for me, typically about 60 pulses a minute.

Both my monitors recorded absolute pressures and pulse rates well. Both showed a lot of scatter, but the pulsation monitor gave results that were on average rather lower that the microphonic one. Conventional hydrostatics told me how to get a falsely low reading—I raised my arm above the level of my heart. This worked. But even with four averaged readings (left arm, right arm, left arm, right arm), neither machine satisfied the scientist in me. The two machines between them implied rather a spread measurement—systolic, ±15 mm Hg; diastolic ±10 mm Hg; pulse rate ±1 per minute. Still, an imprecise blood pressure reading is better than none. I suspect that many people—especially those who implicitly believe LCD numbers—take such machines very seriously.

Occasionally I asked a real doctor to take my blood pressure. The traditional method does not use a microphone or pressure pulsations to detect your pulse; it uses a stethoscope. Probably the doctor knows how to interpret the changing sounds made by your flowing blood and reads the pressures with more insight.

Over the first few years of my hypochondria notebook, most of its entries dealt with medical matters. Thus, like many of us, I occasionally had intermittent low back pain. I noted no useful correlations with my activities, until I had a severe attack after a burst of photography. This required me to take up various awkward postures but not to lean on any support in case I disturbed the optical rig. I now suspect that, for me at any rate, low back pain comes from awkward posture.

I also noted my occasional damned illness, which I blamed on chronically infected nasal sinuses. This distress has always been with me; I even had it in childhood. I have often taken it to the doctor but seldom with any useful result. Unpredictably, my sinuses seemed to flare up. Such an attack, as often recorded in my hypochondria notebook, pushed up my body temperature, gave me dreadful fatigue, and greatly increased my sensitivity to cold. Often I just had to go to bed and keep warm, with one or even two hot water bottles. These attacks could often be cured by a course of antibiotics. I had to go to the doctor for this; for me Septrin (now discontinued) was the most effective. Penicillin V and amoxicillin seemed useful; erythromycin was for me almost useless.

I also wondered about my sensitivity to cold, as triggered by that illness. Was I truly cold or was this a nervous hypersensitivity? I experimented with a contact thermometer. I stuck it on various parts of my skin, to measure my skin temperature directly, and compared it with normal. I failed to reach any clear conclusion, but the idea was worth trying.

As a medical record, the hypochondria notebook justified its whole existence on one supreme occasion. That was the time in 1996 when I got anginal pains—a sense of pressure in the chest, sometimes mutating to a sort of burn. I fought that diagnosis for months and even found a medical paper supporting my interpretation.[1] I wanted it to be an esophageal thing, curable with a drug, or a lung thing, a development of lung problems I had had before. It wasn't, and wasn't.

When the symptoms began in 1996, I visited the university medical library to read up about angina. It took a lot of courage. Angina is a common complaint. It comes from the narrowing of the arteries supplying blood to the heart, by the deposition in them of a fatty layer containing cholesterol. I feared that angina could not be reversed; once the deposit was there, it stayed whatever you did. And yet, I read in one book that maybe 18% of angina victims had a form that diminished if their blood cholesterol was reduced by suitable drugs. Now in my hypochondria notebook I came across a comment dated 1990 that seemed to show I had had anginal pains before. I had not recognized them as such, and they had gone away! Maybe I was one of the fortunate 18% for which anginal deposits can be reduced?

I took my troubles to the medical profession. The critical diagnostic technique is to get a patient to exercise on a treadmill, while recording blood pressure and taking an electrocardiogram (EKG). I was interested in the treadmill and an obvious error in it (fig. 13.1). Furthermore, the technician taking my blood pressure did it badly. Nonetheless, a cardiac specialist took one look at the EKG trace and said, yes: this man has angina.

The diagnosis depressed me greatly. I reacted by trying to get rid of my angina: my hypochondria notebook implied that I had done this once already. I decided to change my diet, reducing the saturated fats from which the body makes cholesterol. Thus I abandoned butter, margarine, cheese, milk, and all fried foods. I gave my doctor regular samples of blood for the standard NHS cholesterol blood test. As a sort of calibration (and also because I was curious about the chemistry of the device), I purchased a cholesterol-measuring gadget from the local pharmacist. I compared its result with the NHS report. The findings agreed fairly well, with the purchased pharmaceutical device giving results a bit lower.

I also considered the "dietary supplements" that many people take. My feeling is that a normal Western diet keeps you safe from almost any deficiency. But in 1996 I reckoned that my antifat campaign, which implied giving up milk products such as butter, might reduce my intake of calcium and magnesium. So I put calcium carbonate (2 g) and magnesium oxide (0.5 g) in my porridge every morning. Both are tasteless white powders. The quantities correspond to the tabulated daily requirements. I have no idea if either substance is efficiently absorbed by the intestinal tract, and I have no reason to imagine that they have done any good. But the experiment was easy enough.

I also considered the virtues of taking exercise more seriously. Some people like exercise. They do it regularly with apparent pleasure. There is even a claim that exercise releases endorphins—pleasure-generating substances—into the system, and this counters depression. (J. B. S. Haldane says that he became addicted to exercise during World War I and had to

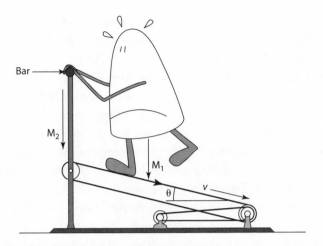

FIGURE 13.1   How to Cheat on a Medical Treadmill

To cheat on a medical treadmill, you simply push down on the bar with your arms. The operator records your full weight M. The machine responds to the weight M_1 you put on the belt but ignores the weight M_2 you put with your arms on the bar. Of course, $M = M_1 + M_2$. The system records your power as $P = M \times g \times v \times \sin\Theta$, where g is the acceleration of gravity, v is the velocity of the belt, and Θ is its angle, but M is your measured full weight. A really desperate patient might push so hard with his arms that the bar took all his weight. He would then merely be pedaling the belt and putting out no power at all. Meanwhile, the treadmill operator would record an entirely fictitious P based on the patient's full weight.

fight for 6 months to conquer the addiction and give it up.[2]) I hate exercise and always have, and I took it up only for my health. So I approached the university medical library with a question in my mind that seems not to have been asked about it. How little exercise can you get away with? I never found a clear answer. But I judged from my reading that to keep in health, daily exercise was too much; weekly exercise too little. Once every 3 days seemed about right. Furthermore, I wanted to extend my exertions till I reached some sort of equilibrium of effort. This might take 20 minutes or more. And there seemed no point in mere mild exercise, like walking or bicycling around. If you were going to the trouble of exercising at all, you needed to stress the system to its limits each time. You thus let the body know what it must be prepared to do.

Odd Notions I Have Played With   189

The exercise I chose was repeatedly climbing the stairs. I had invented it some years ago. It had six advantages. First, it was private: unlike the joggers I could see from my window, or the clients of a gymnasium, I would not be making a fool of myself in front of others. Second, I could make fine adjustments by varying my rate of climb. Third, knowing my weight and rate of climb and the height of the stairs, I could calibrate my efforts exactly in watts. Fourth, I avoided the treadmill error. Fifth, I could take other measurements (e.g., body temperature, pulse, and blood pressure) in the course of the exercise. And sixth, if I felt odd in any way, I could stop at once. I would not have to walk home.

For about 10 years, from 1990 to 2000, I developed this form of exercise. For the first few years it was very sporadic, but after the angina episode I took it more seriously. I got into the habit of warming up at some lesser intensity for 5 minutes first. I extended my main exercise to some 30 minutes every 3 days. My impression was that I hit, and slowly overcame, two limits. The first was my lungs (I got winded). Gradually this limit ceased to trouble me. The second limit was my general system (I got fatigued). I slowly overcame that limit, too. Then I hit a third limit that seemed fundamental, and I never overcame it. I got overheated. Even stripped down to shorts and cooling myself regularly with a wet towel, I still got hot. My body temperature (measured with a medical thermometer) rose too. Gradually my anginal pains, which after my initial experience of them came on regularly when I got into this exercise, went away.

Over the years in which I developed my stair climb, I managed to increase my mechanical output from about 60 watts of power to about 100 watts. I could measure my weight (m, in kilograms), the rise of my stairs (h, in meters), and the time it took me to ascend that rise and come down again (t, in seconds). The power I put out during my climb was then $P = mgh/t$ in watts (g, the acceleration due to gravity, which gives you weight in the first place, is 9.81 meters per second per second). At my most vigorous, this came to about 100 watts mechanical. Guessing my climbing muscles to be 20% efficient, 100 watts mechanical implies 500 watts metabolic. Further assuming that the rest of me (brain, stomach, and so on, and running down the stairs between each climb) was also using up 100 watts metabolic on average, my exercise reached a total metabolic output of some 600 watts.

I would love to have measured my metabolic rate directly, say by measuring the rate at which I converted atmospheric oxygen to carbon dioxide and water vapor. But I never thought of a good way to do this. Instead, my metabolic assessment relied on that 20% guess.

Even so, 30 minutes at 600 metabolic watts is quite creditable for an elderly sedentary intellectual. Bryan Allen, the pilot who pedaled the Gossamer Condor man-powered aircraft, could put out about 1400 metabolic watts for 30 minutes! Furthermore, my 30 minutes at 600 watts is about 300 kilocalories, corresponding to some 37 grams of fat. Eating could easily make it up. So if you merely want to lose weight, eating less is far more efficient than any amount of exercise.

Did my exertions have any beneficial effects? It is hard to say. The anginal pains of 1996 slowly went away; maybe my saturated-fat-free diet helped. My blood pressure and blood cholesterol numbers dropped a bit, and my weight dropped a bit. Then in 2000 I had a brain bleed (which seems to have had no connection with the exercise; it was two days after one) and I gave it all up. The fact that I survived at all, when the brain surgeon feared I was done for, may show that fitness is useful. I still maintain the hypochondria notebook and still regard myself as a biological specimen about which I have intimate and detailed knowledge.

What other hypochondriac jottings are noteworthy? My weight seemed to drift unpredictably all the time. Thus I lost weight during a depression and more than made it up when I recovered. I remain amazed by the extent of that loss—about 9% (fig. 13.2). As far as I can judge, the weight came mainly off my stomach. Hence, perhaps, the male variation in trouser waist-measurement during life.

Temperatures are worth noting, too. My bath water is at 41°C, compared with a cup of hot beverage at 65°C. The difference, 24 degrees Celsius, implies that my skin can tolerate about 24 degrees Celsius more than what it is used to. With skin under clothing at about 17°C, the hottest bath is again 24 degrees Celsius hotter, at 41°C. The inside of my mouth is at 37°C, and 24 degrees Celsius above that is 61°C—not far from hot-tea temperature.

My hot-bath temperature was rather critical. My sense was 41°C was almost too hot, while 37°C was rather tepid. I now muse that those who prefer showers may be able to enjoy a wider range of water-temperatures on body skin. Similarly, those who like sleeping under a duvet may be

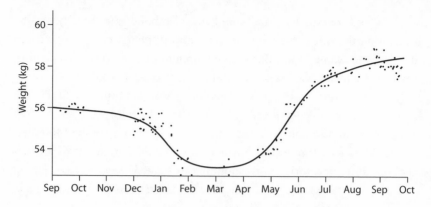

Figure 13.2   Author's Weight during a Depression
The dots are individual weight measurements. Each may be in error; and
the author's weight itself must have varied from day to day—with his water
content, for example. The line smoothes the points. It probably shows how the
author's weight varied over the year of the weighings.

able to tolerate a wider spread of surface temperatures than those who
prefer sheets and blankets. These let you make fine thermal adjustments,
by peeling off a layer.

One personal biological experiment I greatly envy is terribly simple.
At age 32, W. B. Bean made a mark on the base of his fingernail.[3] As the
nail grew, the mark rose; in due course it reached the top end of the nail
and he clipped it off. He noted when he did this and made another mark
at the base of the nail. He kept on doing this. Over his life he thus accumulated an accurate record of how a nail grows (it slowed each winter
and stopped during an attack of mumps). How simple! How elegant! And
I never thought of it!

I suspect that my hypochondria notebook is a worthwhile part of my
idea mechanism. I have somewhere to put personal queries that I would
otherwise forget. It has made me think about matters such as muscular
efficiency and thermal tolerance that I might not have bothered with otherwise. It has stimulated my notions about diet and the human body as
a machine. And it has broadened my range of silly questions. Thus when
I had an EKG once, I asked the technician if she had ever had a patient
with a heart on the right side. She had had two! I then asked her if they
had been left-handed patients. She had not thought to ask.

Tetrachromatics

I was once engaged to devise a TV program or series of programs in which the whole audience would have acted as a scientific sample. Some visual challenge would appear on the screen, and many of the audience would telephone in their responses.

I liked the notion of using a TV audience as a scientific sample. I was perhaps inspired by the "rain on the toilet paper" idea I devised for Yorkshire Television Ltd., some time in the 1970s. Aha! I thought. What a splendid new idea for TV! The plan was for Magnus Pyke to give the challenge at the end of the program. "Do this now!" he was to say. "Don't wait for the ads! Go to the lavatory, take a piece of paper, hold it outside in the rain for 5 seconds, and tell us how many drops you counted!" I had a lot of fun buying all the brands of toilet paper I could. I calibrated them for the drop-catching ability of one sheet and translated it into rate of rainfall. All toilet papers were much the same.

The television producers then had a hard problem. The program was recorded on a Wednesday, for transmission next evening. Would it rain tomorrow? We were phoning up the Bracknell weather experts and American meteorologists with access to satellite data and finally took the risk of including the challenge in the program. It worked! Not only did it rain in the United Kingdom on Thursday evening; there was quite a detailed pattern of rain, with some dry areas. This was before the radar mapping of rain, so our appeal had scientific value. Children all over the country, and many adults too, rushed to the lavatory to do Magnus's bidding. Indeed, YTV gave our data to an Imperial College meteorologist, and one of his students built his MS thesis partly on the data we got that night.

Anyway, musing on that triumph, I had several other ideas for using a TV audience as a scientific sample. My first thoughts were to use TV as a sort of mass medical test. I mused about showing a Snellen optical chart, with rows of letters for the audience to see, or playing audio notes of various frequencies for them to hear. But there were obvious snags. Then I thought of the "subliminal advertisements" we used to be so frightened of. They were to be inserted into a program perhaps as a few frames or even a single frame: but could the concept itself ever work? I liked the idea of testing it on a big audience.

The subliminal notion I was wrestling with may be relevant to the theme of this book. Observational intuition may get information past the Censor and into the RIG without the Observer-Reasoner ever being aware of it, or perhaps after the Observer-Reasoner has discarded or forgotten it (see chapter 1). How could you display such small intuitive signals on TV? And how could you tell if any member of the audience had retained them? My idea at the time was to flash a color on the screen as a single frame and then ask the TV audience if a particular color "tickled their fancy" at that moment. I did not want to ask for people's favorite color as many of us have a fixed favorite.

Another scheme I proposed also related to intuition, this time the physical intuition (see chapter 4) of the TV audience. You would set up some experiment on the screen and ask the audience what would happen. Then you would try it. Yet another was "disobeying the instructions." You would take some commercial product and flagrantly disobey the rules of use. What would happen if you just kept boiling a cabbage, topping up the water as it boiled away? Or if you just ran a car and never changed the oil or just allowed the tires to go bald?

Then there was my tetrachromaticity survey—a search for people who can see in four colors. Color vision features in much modern technology but depends on the fact that most people are trichromats. They see in three colors. Thus a computer screen has three sorts of colored dot: mine has red, green, and blue. Artists mix paint from three primary colors. Photographic color film is sensitive to three colors.

Color-blindness upsets things. Thus some people cannot distinguish between red and green (the great early chemist John Dalton was like this. His type of color-blindness is often known as Daltonism). Some people are monochromats. They seem to see things in shades of gray only. Color-blindness is usually tested for by a special booklet. Its pictures are made up from colored dots. If you have a specific form of color-blindness, you may see a number in a certain picture where someone with normal vision will see a different number or no number at all. I looked at the whole suite of tests when I went to work for a printing-ink company (my vision seemed normal enough for them).

My idea was to screen a whole TV audience for tetrachromaticity. Perhaps a few people are genetically unusual. They are supernormal and can see in four colors. Doctors and opticians will have told them that they see color wrongly, and they may think they are just color-blind. So I began to devise a color chart to detect tetrachromats. It was based on a simple idea. For normal trichromats, blue and yellow are separate colors and are in the rainbow. When you mix them, you get green. There is also a pure green (green is in the rainbow too). A trichromat could not tell any difference between a properly mixed green and a pure green; a tetrachromat could. (Perhaps. I had to guess what might be supernormal in tetrachromatic vision. But there were at least two other pairs of colors I hoped to try in my chart.)

Before I got very far I realized that color TV itself was trichromatic. A TV camera would not see the subtleties of my chart, nor would a color TV receiver reproduce them. I gave up the whole idea. And in the event, I never succeeded in consulting for a TV program that used a whole TV audience as a scientific sample.

Much later, I decided to look at color vision again. If I ever completed it, my chart might perhaps find a place in an art museum. In such a museum, real people look directly at real exhibits, without any TV technology intervening. After a long while, my chart might find a tetrachromat among the patrons.

I have since mused further about a color chart in an art museum. Some people have the strange useful skill of "carrying a color in their head." They can look at a color, go into a shop where the lighting may be quite different, and still choose a fabric or a paper of the right color to match the sample they have in mind. A test in a museum might require patrons to look at a colored subject and go to a distant chart where they try to match it. The successes, and the types of failure, could tell us a lot about the visual capacities and the skills of the public. A large colored object might be easier to match to a large distant patch. Size might matter a lot—it does in some animals. A museum would study a broad range of people, too. They might be older and closer to the human norm than the students who crowd the physiology textbooks.

So I continue to muse about color vision and its variations, both among humans and in animals. Some insects, perhaps, see the ultraviolet. They may see exciting colors in a flower where humans see nothing. Birds, too, may be able to see color in the plumage of other birds, where we see nothing. Conversely, they may be blind to colors that we can see. I have read that you can discover how ants behave in complete darkness, by illuminating an anthill with red light which they cannot see. I once tried flashing a red laser pointer at a fly. It failed to react in any way, and I decided that it simply could not see the red light.

The military have a night vision for guns, by which soldiers can detect infrared military beacons, or people in the dark. I like the idea of inventing a more general "false color" system, perhaps as a pair of binoculars. You might greatly broaden a narrow spectrum of color, so that a small range of blue (say) widened to encompass the whole human range from red to violet. Or you might shift the whole color scale up into the ultraviolet or down into the infrared. Naturalists might greatly appreciate false-color binoculars. Sadly, I fear they would be very complicated. In any case they could not give true tetrachromatic vision.

Down with Molecules!

I have been a chemist all my adult life. Like all other chemists, I have thought in terms of clean compounds, each with a well-defined molecular structure. But a ghost in my head, sent perhaps from my RIG, has imagined quite another chemistry, a parallel one, without molecules at all. Atoms tend to stick together. That's why solids exist in the first place and why glues work. So one might imagine an entirely different sort of solid chemistry, without any molecules.

In the early days chemists made various stuffs, each with a known or at any rate knowable molecular structure. They were happy to see what sorts of substances they could make. Around 1800 John Dalton imagined that the atom, the ultimate particle of any chemical element, had a specific weight. He imagined that each chemical substance was made of "molecules," each an agglomeration of atoms. A pure substance was made up of many identical agglomerations, many identical molecules. Much later, Friedrich August Kekulé (see chapter 1) clarified the idea. He said that when atoms combine into molecules, they do not clump into a random agglomeration but form a precise molecular structure. Each atom in that structure extends a number (its "valency") of bonds, and each bond has to join to another atom. Many bonds have fairly fixed directions in space (the valency of carbon, for example, is often four; when it is, those four bonds are often directed toward to the corners of a somewhat flexible tetrahedron). Other atoms seem pretty satisfied by whatever bonds come their way. On this philosophy, each chemical substance is made of innumerable identical molecules with the same shape. That shape is often complex and three-dimensional. A solid whose identical molecules are aligned in arrays and point in the same direction is a crystal. It has straight edges (see chapter 16).

This theory, explaining so much practical fact, has been accepted. For any compound, known or new, the chemist's ambition is to assign it a specific molecular structure, with each of its atoms tied in place by believable chemical bonds. Indeed, I and a few million other chemists have spent their lives making and exploring chemical compounds and assigning a molecular structure to each of them. Well over twenty million chemical compounds are now known, and all chemists have accepted the molecular idea. A molecule may dissolve in a solvent such as water, where even without stirring it wanders around. Almost all chemical reactions require two reagents each to be dissolved in a solvent. When the solutions are mixed, the wandering molecules can get at each other. They may then exchange atoms or interact in other chemical ways.

Yet the dissolving of a solid in a liquid, forming a solution, is strange. Heat is often given out, showing that some sort of chemical reaction is occurring. Yet, while a traditional chemical reaction has fixed proportions and rejects a surplus of either ingredient, a solution can be made up in any proportion. Claude-Louis Berthollet (1748–1822) regarded solutions as true chemical compounds of undefined composition. As far as I know, he did not try cooling them to solids and examining the properties of each solid.

This thumbnail sketch of chemistry tells briefly how the science has developed since its early days. Yet the ghost in my head does not keep quiet. Can there be a substance not composed of molecules? "Amorphous carbon" may be one; it could be microcrystalline graphite or some undefined agglomeration of carbon atoms. And solid metals seem just to be atoms stuck together. They may be made of crystals, but the atoms within a metal crystal may simply cling together. Chemists have called that clinging a "metallic bond" but have not given it any particular spatial direction or numerical valency. Some chemists have even made "metal cluster" compounds in which a clump of metal atoms is joined by some sort of chemical bond to a more conventional molecular residue.

Furthermore, metallurgy has not been absorbed by chemistry. You can melt a metal to a liquid (a very traditional element of technology) and then cast it into a statue or an item of goods. You can often even boil it to a vapor. The vapor consists typically of single or double atoms and tells you nothing about the liquid or the solid. Solid metals are often very strong, especially in the form of wires. And even if deformed, a solid metal tends to reform the bonds between its atoms. A metal often bends where other solids would crack or shatter.

So I was very happy to come across a *Chemistry and Industry* of 1964. It had a sort of throwaway line. H. Mackay stated that the gas phosphine partly condensed to an odd sort of solid that seemed not to have an exact composition.[4] It seemed a sort of lattice of phosphorus atoms. Where spare chemical bonds protruded from the mess, they were terminated by hydrogen atoms. I felt that this non-molecular solid was a mighty step forward, and Daedalus leaped on it.[5] He expounded a chemically outrageous non-molecular condition, and I am now extending his argument. To him, the solid was not a polymer but a "randomer." He set DREADCO chemists to discover what systems of atoms form randomers (fig. 13.3).

My chemical instinct is that the best randomers will come from elements whose valency can take many values. Phosphorus is the basis of the only known randomer, but nitrogen seems feasible, and so does sulfur. So does carbon, whose valency can be four, three or two, provided that some other atom satisfies the other end of the bond. Indeed, I have mused that amorphous carbon may be a natural randomer. I have never been convinced by the claims that it is a just a microcrystalline form of graphite. My experience with charcoal filtration (see chapter 7) may have biased me, of course. Silicon is possible too; but most of the other elements with several valencies are metals. Even in their conventional chemical compounds, metals often show several valencies (iron, for example, forms the ferrous series of compounds, in which its valency is 2; and the ferric series, in which its valency is 3).

Any randomer will need a certain number of hydrogen or halogen atoms to mop up unsatisfied valencies. Some randomers will break into

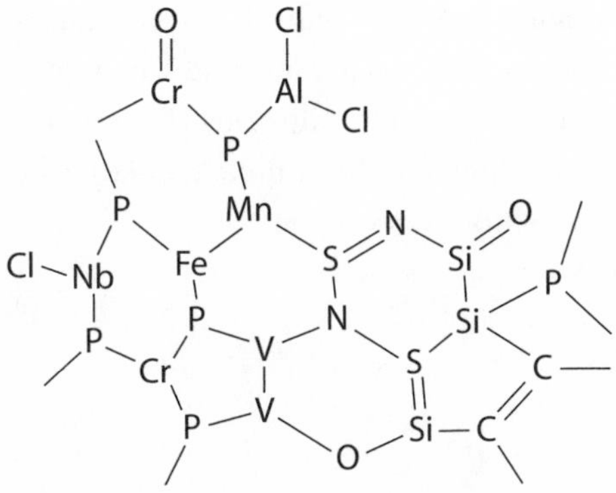

FIGURE 13.3   A Guess at a Randomer
A randomer is a word I coined for a random, planless, collection of atoms
bound together by halfway-feasible chemical bonds. It does not have a struc-
tured molecule. It is not chemically well-known; indeed only one example has
been described. It would be a new solid, and I imagine it extending indefinitely
in all dimensions, rather like a chunk of metal.

small molecules; they will decompose slowly or rapidly. But some will be
stable. A stable randomer will have lots of satisfied valencies and well-
made chemical bonds. It will be a solid, but without any defined or re-
producible composition. It will not make traditional chemical sense. But
I remember the metals and the metal alloys, made simply by mixing the
molten metals in the correct proportions. They also make no chemical
sense, but the cooled result is often extraordinarily useful. Stainless steel
was invented by accident. And think of that wonderful cutting metal,
high-speed steel! You melt steel and add a few percent of tungsten, molybdenum, vanadium, and cobalt, each in a precise proportion. The cooled
melt yields a very hard and resistive metal, much used for drill bits, milling cutters, lathe tools, and similar machine parts.

What use will randomers be? I am quite in the dark. I'd love them to
be as hard and durable as metals, doing the sort of thing that metals are so
good at but much lighter for the strength. These days, we are as interested
in rigidity as in sheer strength (that's why carbon fiber is important), and
maybe randomers will be rigid. Or maybe their electrical properties will

be intriguing, though I fear they will just be insulators, like most polymers. But whatever property you are interested in, you will be able to edge it along in the desired direction by adjusting the initial composition of the randomer. Like a metallic alloy but unlike a chemical compound, it will have no fixed composition to stop you.

14

Literary Information

I have been writing all my life; indeed, my nineteen hundred Daedalus columns are a major oeuvre in themselves. I have written lots of bigger articles too. So I have a deep feeling for the language. I am especially sensitive to writing style. The bare text may reveal the author's Observer-Reasoner, but the style says something about the unconscious mind behind it. A perceptive critic might even guess at the author. I'd hate to try it, especially on a scientific topic. Scientific papers are almost deliberately complex and difficult. But even a scientist sometimes uses the language appealingly. Thus J. E. Gordon, on how wood creeps under load, says "the rather badly stuck hydroxyls take advantage of the changes in moisture and temperature to shuffle away from their responsibilities."[1] I wish I could write like that! Gordon's style is widely renowned.

Literary Styles

Without scientific constraints, language can be much freer. For one professional paper I tried to be more open—that was my scientific paper on the theory of the bicycle (see chapter 5). As a chemist, I felt able to write more freely in a physics journal. And readers seemed to like it. It was even reprinted—the only scientific paper I know to have been thus honored! At least three times I have been asked to translate the scientific diction of some medical diagnosis or pronouncement, so that it made sense to those concerned. On another occasion I translated a scientific paper into ordinary language.[2] I caused a lot of fuss. (That paper was about handedness. It began, "It is common knowledge that contemporary man prefers to use his right hand when performing unimanual tasks." My translation was "Most people nowadays are right-handed." And so on.) Freeman Dyson of *The Physical Review* has denied that complex scientific prose merely reflects the subject.[3] He says that *The Physical Review* rejects most papers because it is possible to understand them. "Those which are impossible to understand are usually published."

A scientist depends for his career on getting his papers published. He writes even if he has no skill in the art and is unable to put his findings into a style that editors can accept. Many scientific eccentrics use the language very badly. Stefan Marinov, for example, could not even get his papers published. He used to buy space in *New Scientist* or *Nature*, for example, and put forward his views of physics in the form of paid advertisements.[4] This freed him from editorial control but denied him a place in the scientific literature. Advertisements seldom have a page number to refer to and do not go in the bound library volumes of a journal. Yet even a scientific advertisement should hope to convince its readers. Marinov's wild diatribes tended instead to antagonize them. He would denounce his opponents as jellyfish or blockheads, give findings without experimental detail, and claim that his arguments were obvious to any child.

I have met many scientific eccentrics, in potential papers, in correspondence, and even in person. *Speculations in Science and Technology* was a copious source—I was on its Editorial Board. William Honig, the editor, has written about the crazy papers it received.[5] And in just one of those papers I found a good argument—the idea that random ideas arise in the brain as a result of the decay of radioactive atoms.[6] In the moment of decay these put out high-speed electrons, and sometimes fire a neuron. Is such an unpredictable event part of mental creativity? If so, it has a place in this book. I have asked astronauts if they have had any sudden idea in space, triggered perhaps by spatial cosmic rays, or the Van Allen radiation belts. (They said no.) But during an x-ray brain scan I once had a sudden unprovoked memory. Did an x-ray photon trigger a cell in my brain? Daedalus has used this notion.[7]

Underwood Dudley's paper "What to Do When the Trisector Comes" deals with those who claim to be able to trisect an angle with ruler and compass only (something it is easy to show is mathematically impossible).[8] His book explains how to calm such cranks or at least get rid of them and their letters.[9] My own experience has been with Daedalian letters and perpetual-motion cranks.

In the course of my curious scientific career I made several fake perpetual-motion machines, the true power source of which was concealed. To my astonishment I discovered a whole group of crackpots who were trying to make perpetual-motion machines for real! I told them that I was a fraud and a charlatan; they accused me of lying to protect my secret. One of them actually said in my hearing the archetypal words, "All I want is stronger magnets."

Of course, perpetual motion is quite impossible. Any engine needs a supply of some energetic fuel and stops when this is exhausted. The laws of thermodynamics are a very firm part of science and show that there is a fixed amount of usable energy in the world. (At present we are hunting for new sources of energy—see chapter 15.) The French Royal Academy of Sciences resolved to accept no more claims of perpetual motion in 1775; the more tolerant British Patent Office made the same decision in 1937. No scientist doubts the principle; but that has not stopped the crackpots or the charlatans—I am one of the latter.

I fear I laid myself open to this trouble. By being Daedalus, part of *New Scientist* and the University of Newcastle upon Tyne, and by making and exhibiting strange objects, I sailed closer to the edge than reasonable scientists do. One ploy I tried, which definitely did not work, was to put each crank in touch with another. I would refer to each as an "expert in the field" and hope that both would then write their painful missives to each other. It never worked; both then wrote to me. Each probably recognized the other as just a crank, whereas I seemed to be a genuine scientist. Each wanted me to support his claims!

A crank letter used to be immediately recognizable. It was typed, typically with a clapped out black-and-red typewriter ribbon that showed odd patches of red on parts of some letters. I suspect that from long use the ribbon was imperfect and had the wrong width over some of its length. The nut letter went from edge to edge of the paper, and from the top to the bottom, with no margins, as if paper (like typing ribbon) were too precious to waste. There was no underlining (I suspect that a nut in the full flow of composition does not to want to move the typewriter carriage back for underlining), but additions were scrawled around the typescript in black ballpoint pen. When I read the thing, I found that its style was very strange. It was both odd and urgent, as if I were already familiar with the topic, but needed an intellectual push to be convinced. Often it was one long sentence. Strangely, the whole thing was sometimes a carbon copy. Only men seemed to write such letters: I have only received one from a woman.

My reaction to such missives was fairly standard. I would write one courteous reply—after all, I was usually acting as a magazine author and did not want to annoy a reader by my failure to react. I had a standard phrase—"there may be something in what you say"—which has a mollifying effect out of all proportion to what it actually gives away. But I would not get involved in long correspondences.

There were several types of nut letter. I have heard of a bishop who received many nut letters and judged them at once by one characteristic—lined paper. I have not received the sort of letter in green ink written as if by some hate-filled madman gripping the pen with both hands. Neither have I received the ultimate nut letter that tells me how wonderful I am and ends, "Please excuse the crayon, but they won't let us have sharp things in here."

I fear word-processing software and e-mail have killed the nut letter. You now have to read the thing to authenticate it. I doubt that the cranks have gone away—we can all get obsessed by a problem. But they may be a little harder to spot. My friends in *Nature* tell me that they still get "inappropriate submissions," often hand-written. So the art is not quite dead.

Linguistic style is as variable as authors. Samuel Butler once hoped that he had no style—that his writing was just simple and straightforward. But we all have a style, or perhaps several, depending on what we are trying to do, and to whom. Primo Levi has said that clear and concise writing is part of the commercial contract between an author and a reader. I strongly approve of the sentiment, but it still leaves many stylistic options open. I can only say that easy reading is damned hard writing.

The style furthest removed from scientific English is that of poetry. Much poetry does not even make sense! I guess that good poetry touches the unconscious mind and is not aimed at the Observer-Reasoner at all. I encountered the peculiarities of poetry forcefully when I came across Edward Wheeler at Imperial College. I could write prose as well as he could, but he could write songs and poetry in a way that I never managed. I was annoyed and saw his skills as a sort of special personal ability. Later

I decided that his RIG was always on the lookout for snatches he could use and played around with them. It was very talented. In due course it pushed the products upstairs; from the pieces Edward was often able to create a song or a poem. This is no sort of explanation of the poetic gift, of course. Some people just have it. Others are very literal and technically minded. Few people combine both skills. As a result, there are few technical poems; I am glad to note such as have come my way. I have even tried to add to their number. Here, for example, is a poem I wrote years ago on the physical constants:

As God devised the Universe, there grew at His fiat,
The table of physical constants as He pulled them out of His hat.
Before He said "Let there be light!" He chose, to meet His need,
The arbitrary constant c to represent its speed.
And in His new creation, He wanted there to be
A law of gravitation; and so selected G.

He chose the electronic charge, in His creative way,
Defined the particle masses, invented h and k,
Till with a set of numbers, He'd totally constrained
A single mighty Universe, and all that it contained!

And in their implications, according to God's plan,
The fundamental constants define the creature Man,
To whom He gives this warning—that time may now be short,
That even He will weary of the things that He has wrought,
And soon may sound the trumpet, for that awful moment when
The fundamental constants will be shuffled and dealt again!

It would be a final triumph of science to show that the fundamental constants had to have exactly their observed values. At present they just seem pulled out a hat—we can only measure them. Many physicists have imagined universes in which the constants have other values (I discuss other universes in chapter 7). Such universes often seem very boring. Stars and matter never form, and nothing ever happens. By contrast, our own universe has some sort of exciting detail on every scale of space and time. Maybe God chose the physical constants for maximum entertainment value. The evolution of life, of course, is one of the major entertainments.

Names

Giving things names is a fundamental problem in any language. One of the its purest forms was faced by the chemists who discovered so many new chemical elements. None was named by its discoverer after himself—Primo Levi saw this as an act of amazing modesty. But four (Germanium, Polonium, Francium, and Americium) were named after nations.

Compounds of the elements were even harder to name; ultimately systematic chemical nomenclature was invented and imposed. Trade names pose a still more poetic challenge: they face competition. The greatest of them all is perhaps Kodak, for the photographic film and many associated objects. It means nothing in any language but sticks in the mind. It belongs to the Eastman Chemical company. I reckon that the rival chemical company Du Pont suffers from severe Kodak-envy: nearly all its trade names also have two syllables with the stress on the first (Nylon, Lycra, Lucite, Corfam, Delrin, Teflon, Mylar, and many others). I also admire an American Army vehicle of World War II. The name Jeep was not dreamed up by advertisers at all and yet has great poetic power. (It was Army slang for "General-Purpose Vehicle.") I know only a few products named after their human inventors; their commercial record is patchy. Parkesine, a Victorian form of celluloid invented by Alexander Parkes, failed. But Bakelite, first of the thermoset resins, invented in 1906 by the chemical innovator Louis Baekeland, succeeded. It is still in use.

Around 1900, perhaps, you could make a trade name just by putting an O on a simple root—Jell-O, Bisto, Glaxo, Silvo, Meccano, Oxo, and so on. The humorist John Morton grew up with such products and made fun of them mercilessly when he took up journalism. More recently, short brief commercial titles have become popular. The National Biscuit Company became Nabisco, Associated Dairies became Asda, the Atlantic-Richfield Company became Arco, and the American Oil Company became Amoco (later British Petroleum swallowed both, and changed its name to BP). Yet one snappy name change failed. In the early 2000s the Post Office Ltd (an umbrella organization which covers the Post Office, Parcel Force, and Royal Mail) spent about £2,000,000 changing its name to Consignia. The British public, however, preferred the old-established office. It had the unique humility to change back.

Some names are funny. Thus Edward Wheeler invented that crucial chunk of the Daedalus column, the company DREADCO. I salute it in chapter 2. (He also came up with the Greek philosopher Mediocrates.) A funny name may speak to the unconscious mind as a sort of condensed poem. The novelist Charles Dickens is justly famous for his names—Ebenezer Scrooge, Uriah Heep, Mrs. Gamp, Mr. Pickwick, and so on. In their day they must have had strong humorous power. But public taste has moved on, and most of us no longer feel the humor of Dickensian names. They are merely odd.

I divide funny names into (at least) two divisions. The first is funny because, as pronounced, it suggests a meaning, appropriate or undermining. An example is John Morton's foreign potentate, the Khur of Khashdoun. "Khur," as well as recalling Kaiser and other kingly names, also suggests a dog; Khashdoun could be a town or a kingdom, but also implies brutally direct financial dealing. Another type of funny name depends on more abstract twitting of the syllabic soup that makes up a language. A good example is Michael Wharton's Sadcake Park. I do not know why this is funny, though the concept of a sad cake has some sort of appeal (I think of a sponge cake that has gone flat).

Morton, who was Beachcomber of the UK newspaper *The Daily Express*, and Wharton, who was Peter Simple of the UK newspaper *The Daily Telegraph*, are now both dead. I do not know who currently has their "funny-name" crown. But I can still applaud their product. My primary examples of Morton's funny names are foreign, but with some clear English meaning. They include the Russian ballet dancers Sonia Tumbelova and Serge Trouserin, the Italian singer Emilia Rustiguzzi and the Italian violinists Screechi and Scracchi. Morton also invented the film stars Trivia Tansy and Dawn Kedgeree and the Scottish nobleman the Laird of Shrillwillie. His foreign achievements included the Khur of Khashdoun (mentioned above), the country Filthistan, and its town Thurralibad.

Michael Wharton went in for syllabic twittery. His industrial UK towns included Stretchford, Nerdley, and Soup Hales. (The first of these contained beautiful, sex-maniac-haunted Sadcake Park, and also the road Numb Lane.) He invented General Nidgett of the army tailoring corps (renowned at el Alamein), H. Bonington Jagworth of the motorist's liberation front and the entertainer Marylou Ogreburg of the Bread and

Marmite Street Theatre Dance Group Collective. He also invented a ship, the *Miss Minnie Baldbrush.*

Both Morton and Wharton could spread into the specialism of the other. Thus Wharton invented Don Binliner of the Stretchford University outreach team, the Rev. Spaceley Trellis of Stretchford Cathedral, the hopeless goalkeeper Albert Rasp, and the unpopular Labour MP Arthur Grudge. Morton invented the Scottish nobleman the Macaroon of Macaroon, and La Belle Zaboula of Wugwell's Circus. He also created firms of solicitors with names such as Oomes, Spickmarl, Twosafrock, Knickerstick, and Wallow-wallow. He invented the products Snibbo and Flubbo, and the soft drink Poopsi-Boola. His legal celebrities Mr Honeyweather Gooseboote, Mr Tinklebury Snapdriver and Justice Cocklecarrot, are widely renowned. Morton may also have invented the African region of Zimbabwe, but if so, it came true on him.

In an intriguing inversion of the art form, Paul Jennings detected subtle meanings in the names of ordinary British towns. Thus the state of being not exactly ill, but certainly a bit below par, he called "being Wembley." Someone who has made a big blunder has committed a Cromer. Jennings also claimed that Pewsey is a sort of narrow-minded religiosity and reckoned that a Kenilworth is a trifling or beggarly amount.

The BBC Radio program *Round the Horne* also went in for funny names, but its sexual subtext leaked into them only indirectly. Thus it had a persistent correspondent, J. Peasemold Gruntfuttock, and a resident songster, Rambling Sid Rumpo. By contrast, the writers behind the BBC TV cartoon program Captain Pugwash (a pirate) tried to slip real sexual implications past the BBC censor. Master Bates and Seaman Staines did this, perhaps, and so did Pugwash's nautical command to "Grease the Buttocks!"

A Personal Database

Many times in this book I have advocated noticing things and remembering them. As an extension of this notion, I have built up over life a personal database of thoughts, quotations, and references.

That database started by accident. At the age of about 20 I came across an article in the Science Museum library and could never find it again. I decided that I would not let this happen a second time. I began to keep a copy or a reference to anything that interested me. I noted some information because it was relevant to a task of mine—an article or a paper, perhaps. I kept other stuff just because it seemed interesting. All sorts of things went into my files—papers, references, scraps of information, ideas for arguments, observations, jokes, odd facts, newspaper cuttings, remarks, queries, anything. The stuff might be in any form, too—a copy, a note, or a code for the item in a library. Incidentally, I salute the photocopier. It developed in my lifetime and made it much easier to record and store information. I have come across the claim that everyone should know "something about everything, and everything about something." I like it. Again, knowing things and knowing about things, having facts and notions in your mind for the RIG to play with, matters!

My own heap of information grew rapidly. At first I kept my items in random order in accordion files. But I soon felt the need to devise an order. The whole thing had grown so big I could not remember what I had got; and I did not want to plough through all that stuff to see if I had something useful. Many libraries use the Dewey decimal system of classification, which groups all topics into a thousand general ones. If needed, any one topic can be divided indefinitely to cover finer detail. But Dewey seemed a bit daunting and formal for me.

After one big false start (which was far too complicated) I divided my collection, not into a thousand basic categories but just nine. These were (1) physics, (2) chemistry, (3) astronomy and geology, (4) biology, (5) practical technology, (6) theoretical and computational technology, (7) intellectual and aesthetic, (8) social, and (9) individual. I soon found this system rather coarse. Chemical engineering, for example, might be 2, chemistry, or 5, practical technology. So I added a bit of complexity; I allowed any topic to have two digits. Thus chemical engineering was 25 (I might have had 52 as well; but this gave too much trouble).

My final classification had nine single-digit items and 36 two-digit items, a total of 45 divisions. It seemed to divide my interests fairly evenly. I did not find everything winding up in one category. Indeed, the least popular category still had about one-fifth as many items as the norm, while the most popular category had about five times. So the scheme seemed to span my interests, my "universe of discourse" fairly well. Others, with a different universe, would classify things differently. Even for me, some interests did not fit. Thus I have another file, of images and photographs. Those 45 divisions do not fit it at all well.

During my first few years at Newcastle, personal computers began to be sensible domestic objects to own and use. So I began to transfer my data to a computer database. I played with several database programs, to avoid troublesome space restrictions or annoying formatting. And I liked the idea of using the same database program for other tasks, such as storing Daedalus columns. I finally settled on Blackwell's "Idealist" program. This was a mistake. Bekon Marketing later took it over, failed to support it, and now does not exist. But there are (at least) five other programs that could do the job. I called my collection of data "Source." I later transferred it, and the Daedalus record, to Microsoft's Access.

Transferring my data to the program was a huge task. But computerization transformed it. At present, the Source database has about six thousand entries. Each may be a one-line remark, a multi-page essay, or anything between. My rule is this: "If in doubt, put it in." If an entry refers to something else, I include an exact reference, so that I can find it in a library. If it tells me that there is a relevant paper in my filing cabinet, it says where. My file of papers, now three drawers of a four-drawer filing cabinet, is the descendant of those early files. One day I may scan the stored documents onto a computer disk, but for the moment they stay on paper. Presented with any query or notion, my first action is to look in Source.

One advantage of my Source data is its antiquity. It has a lot of old stuff that seemed new and exciting when I recorded it. If you just pass something by, you might as well never have encountered it. Once recorded, it retains some sort of personal appeal—even if it gets outdated, or if you later decide that it was nonsense.

I have built up this heap of information all my adult life; and I recommend the slog. I suspect that my RIG kept tabs on what I had got. I even have the impression that I know some of the stuff. And looking out for new material may have kept me mentally active. A small amount—perhaps 1%—has later even come in useful. Much of this book, for example, is based on clues in Source.

Often I recall a brief phrase, or even a single word, of some little saying or strange comment. With Source I can often track it down. At intervals I conduct the boring but necessary task of copying Source onto separate computer disks. I keep these in a separate location. I hope they are secure. I have built Source up over my whole life and would hate to lose it now. Some people have built up a similar "personal database" of scrap notes, which has been stolen or destroyed. They have mourned its loss. But many other personal databases must have survived. Build one—it should help the RIG!

Has the Internet made Source obsolete? Probably not. The Internet often fails to answer the sort of question I want to ask—or more likely, I am not good at Internet searches. Thus for this book I once wanted to know the efficiency of a refrigerator (see chapter 15). I tried to look up the matter on the Internet. I learned a lot about the energy-rating of new refrigerators, their low electrical consumption compared with earlier models, the need to keep their insides clear of ice and their outsides free of dust—but actual efficiency? Watts of cooling compared with watts of electric power consumed? Not given. I failed to find anything useful, and the claim in the next chapter is my own guess. A 1920 physics textbook says that then current machines were about 300% to 700% efficient.[10] (I take the lower estimate.) The information I wanted is probably somewhere in the Internet, but well hidden.

15

Inventions We Need but Don't Have

We modern humans have endless wonderful skills that would seem magical to earlier societies. Yet to my mind, we are far from constructing a stable, sustainable, technically advanced civilization, accepted by all its citizens and not threatening to the global environment. To do that, and yet to stay within the limits revealed by science, we need real technical creativity in many new directions. I hope to encourage that creativity. So from a position of Olympian ignorance, I here scan the whole technical field for the big things that are missing. In this light, our inabilities seem endless; it's hard to know when to stop. But of all the inventions we need (but don't have), I will start with a big one.

Energy

There are perhaps four problems with energy, and I reckon that we are making progress with one of them. First, we must make it on a large scale; second, we must store it on that scale; third (with which we are making some inventive progress), we must provide it neatly in small amounts. The fourth problem is being efficient with it.

Large-scale energy generation seems bad and getting worse. We get energy by burning the combustible portions of the planet (coal, oil, and latterly gas). That combustion releases carbon dioxide, which may enhance global warming.

Our gestures toward "renewable" sources of energy, such as wind and solar power, have only one big success: hydroelectricity. Usually the dam it needs has to coincide with some other use we have for the water. Nuclear power stations leave a ghastly residue of radioactive waste, which we cannot clean up (but see chapter 16). And the human fear of radioactivity is deep rooted (see chapter 16).

All these schemes are big. They feed a big expensive and wasteful electrical grid. It is amazingly reliable, and we take it entirely for granted. In the developed world, it delivers energy at about $0.1 per kilowatt-hour. During the 1970s, several big "renewable energy" schemes were planned. One of them, inevitably American, proposed a big energy satellite in orbit (fig. 15.1), as discussed by R. A. Herendeen.[1] Hundreds of meters across, it could concentrate thousands of megawatts of solar light energy on a silicon photoelectric receiver. Power circuitry on the satellite would convert this to microwaves, which would be sent down as a beam to an aerial farm on Earth. The farm would convert that power to electricity for the national grid.

This project had two main advertised advantages. First, with an inclined orbit the satellite could be in sunshine all the time. It could collect solar light and deliver it to Earth as microwaves even at night. Second, since microwaves can penetrate cloud cover, it would work in cloudy weather. There were two further, unadvertised, advantages. First, it was vastly expensive, and thus added to the big costly infrastructure controlled by the United States federal bureaucracy. Second, it was a weapon. By aiming the microwave beam at a city (you would send a simple order to the satellite), you could overheat that city and maybe ignite parts of it. It is perhaps significant that with the ending of the Cold War, the project is not fervently advocated.

Even so, solar power is one of the renewables worth working on. Daedalus, inevitably, has mused about it.[2] In the form of photosynthesis, invented and evolved by green plants, it drives all life. And in 1954 the legendary Bell Telephone Laboratories (see chapter 2) invented the silicon photocell to capture it directly and turn it into electricity. I have been waiting 50 years, in vain, to see silicon solar energy get really cheap and available. My fantasy is a cheap photoelectric roof tile from which to get domestic electricity. We are still not there. Only in the space industry, where huge costs are the norm, are silicon photocells important. I like the idea of a thermal renewable, the heat of the Earth. Indeed, geothermal heat and even geo-electricity are being worked on but are not much dis-

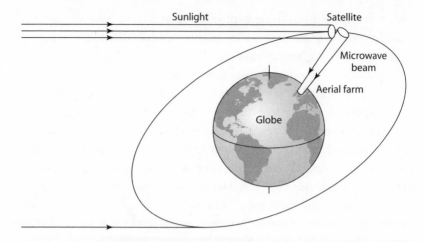

FIGURE 15.1 Proposed Energy Satellite
An energy satellite orbits the terrestrial globe. It intercepts a shaft of sunlight, converts it to a microwave beam, and radiates it down to an aerial farm. This creation was a renewable energy idea proposed in the 1970s.

cussed.[3] Most of our energy still comes from traditional thermal engines, burning the combustible portions of the planet.

The efficiency of such engines is 5–40%; the rest of the energy of the fuel appears as "waste heat." Hence the big cooling towers of landlocked power stations. A Czech power station once disposed of its unwanted heat by delivering it to a nearby town. The result was free compulsory central heating, a socialist benefit indeed! The notion did not spread to capitalist countries or even to other socialist ones. The generation of British electricity wastes more heat than British Gas sells. I discuss thermal limits below.

Nonthermal motors, such as electrical ones, or ones that use falling water or (as in living muscles) chemical changes, escape thermal limitations. They can be 80% or 90% efficient or even more. Our muscles are not thermal engines. They work at body temperature; they react blood glucose with dissolved oxygen. Somehow they convert chemical energy directly to mechanical work, at body heat. This astonishing chemical triumph is far beyond any human technology. Yet muscles are only about 20% efficient; they are not dramatically better than thermal engines. They "waste" about 80% of their chemical energy as heat, for which the body has a use. I suspect that a muscle could be far more than 20% efficient if the body did not welcome its heat.

Again, we cannot store much energy. The renewables, such as wind power, wave power, and solar energy, are notoriously intermittent; we have to fill in the gaps. In particular, solar power is only available during the day; but we need energy at night, too. We cannot store bulk light, nor the energy to make it. Gaston Planté's 150-year-old lead-acid accumulator is still the energy reserve in modern cars (to be fair, clever design lets it kick out for a few seconds the 4 kilowatts or so that starts the main engine). That currently fashionable hybrid car, the Toyota Prius, has a nickel metal hydride battery that weighs 35 kilograms, captures some of the energy usually wasted in braking, and stores about 5 million Joules of energy. By contrast, 35 kilograms of gasoline stores about 1,600 million Joules, and 35 kilograms of animal fat stores about 1,200 million Joules.

So let us imagine an ideal rechargeable battery. Weight for weight, it would store about as much energy as fossil fuel or animal fat, about 300 times as much energy as existing batteries. It would be pretty explosive, but then so is gas—at least gas mixed with air to burn it. That battery would presumably be a fuel cell, and would breathe free air, perhaps 20 times the weight of its stored fuel.

If it existed, such a battery would transform life. It would need no electrical connection and emit no nasty exhaust. It could drive power tools, lawnmowers, vacuum cleaners, wheelchairs, a domestic robot, the Segway walking device, even Clive Sinclair's C5 electric vehicle. We need it!

Our one mighty success is the small rechargeable battery, universally recharged from the electrical grid. It drives mobile telephones and mobile laptop computers, which wonderfully augment our ways of holding and transporting information. But we still have to lug people and stuff around in the old way.

For mobile objects outside (vehicles and aircraft) we burn oil, mainly as gasoline, and almost universally in internal-combustion engines. The light, powerful internal-combustion engine is a triumph of engineering. As a thermal engine, it is only 5–40% efficient and you have to start it. Not until the 1960s was the accumulator battery so reliable that car designers felt comfortable with an engine that could not be started by a handle, even in an emergency. Alec Issigonis took that bold step with the

British Austin Mini. His novel transverse-engine design is now a standard for motor cars.

We waste a lot of energy keeping ourselves warm, too. The electric storage heater uses high-class grid electricity to heat a mass of bricks. It probably made sense as a way of disposing of nuclear power at night (nuclear power stations run best all the time, even during the night when demand is low).

How to improve domestic heating? Since the housing stock now largely exists, we need some sort of retrofitted technology. Hence the schemes for insulating attics and filling cavity walls with foam insulation. One useful modification would be a really clear, really insulating, glass for windows. Glass is what it is. Too bad. Michael Faraday wasted much of his genius trying to invent new sorts and failed. Ideally, we want a glass to let light in freely but not to let heat out. My RIG has imagined a "light-rectifier" that lets radiation one way through glass but reflects it the other way; but I cannot anchor that dream. I can, however, imagine a glass with tiny holes much smaller than a light-wavelength. Light, but not much heat, would go through it. It would make highly heat-insulating windows and wonderful lenses too. Sadly, we cannot make anything like it.

Indeed, modern optics has far to go. We also waste a lot of power in that truly important recent invention—artificial light. How dreary the northern winters must have been, when even the candle had its magic! But a tungsten bulb uses electricity to make light at maybe 0.5% efficiency. "Energy saving" fluorescent lamps push this up to 2% or 3%. One day, but not yet, the light-emitting diode may transform the field by operating at 60% or more. Daedalus, of course, solved the whole problem long ago (in fact in 1973). His photonic plaster, Phaster, absorbed light during the day and gave it out at night—a lovely idea (see chapter 12). But as so often with him, a sound principle and persuasive examples still somehow failed to build a new technology.[4]

Further Energetic Notions

THE CARNOT THEORY OF HEAT ENGINES

A heat engine works by taking heat at a high temperature, turning some of its energy into mechanical work, and releasing the rest at a lower

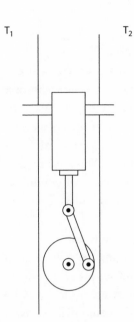

FIGURE 15.2 Carnot's Theory of the Steam Engine

Any heat engine takes heat from a boiler at T_1 and exhausts it to a cooler condenser at T_2. Carnot was familiar with the steam engine, but his theory applies to any heat engine. It even works in reverse for a refrigerator pumping heat from a cold object T_2 to a hotter one at T_1.

temperature. It is rather like a water wheel, which takes water at a height and releases it at a lower elevation. This was a common analogy until, in the early 1800s, Nicolas Carnot deduced the true laws of thermal engines. Their maximum efficiency was, he said,

$$\text{Efficiency} = (T_{boiler} - T_{condenser})/(T_{boiler})$$

Watt's separate condenser for the steam engine (see chapter 1) made that exit temperature ($T_{condenser}$) clear and important. The Ts are on Kelvin's "absolute thermodynamic temperature scale," Centigrade plus 273 (fig. 15.2). Designers of thermal engines have therefore always tried to increase the operating temperature of the boiler (a flame in the internal-combustion engine) and to reduce that of the condenser. Thirty percent efficiency is good. Even when you have got your energy, usually in the form of a forcefully rotating shaft, you may have to tolerate a lot of waste getting it where you want it (as through a complex electricity grid) or doing what you want with it (such as driving a vehicle against the resisting air by turning the wheels at the right sort of speed).

A NOTE ON REFRIGERATORS AND AIR CONDITIONING

The Carnot principle can work in reverse, when it implies a useful gain. Thus in a refrigerator, a "heat pump" driven by a motor can take about 50 watts of electrical energy and use it to pump maybe 150 watts of heat from inside the cabinet to outside it. The inside of the refrigerator cools down and the kitchen warms up. Sadly, the refrigerator cabinet is usually badly insulated, and kitchen heat leaks back in. You have to keep the refrigerator switched on. On a somewhat larger scale, air conditioning can use the same principle to keep us usefully cool; but you have to avoid air leaks!

If we were clever enough, we could use air conditioning to warm ourselves. Imagine outside air at 10°C and our wanting to be 30 degrees warmer at 40°C. A perfect Carnot heat pump could simply take the outside heat and pump it inside. It would cool the outside a bit, but who cares? Furthermore, each watt of mechanical work would deliver about 10 watts of inside heat. But we still find it easier to warm ourselves by burning something.

A NOTE ON GLOBAL WARMING

Burning organic matter for energy, or just to get rid of it, releases carbon dioxide into the air. This is alleged to cause global warming, by retaining extra incoming sunlight. To counter it, you might reduce the solar input by about 1%. I imagine a big reflector in space, in front of the sun, acting like a little permanent eclipse. The leading space-competent nation is America. It should love to design and launch that sunshade and let Americans go on burning oil freely. You would need millions of tons of reflective material, and it would be hard to keep in place. My RIG imagines another way out—a spread of Earth satellites say 100 kilometers up, each a big reflective "flag" of thin aluminized polyester. The satellites would create a succession of little eclipses rather than one permanent one. A neater technical fix for global warming might be to deploy unmanned ships in the seas, each blowing sea water droplets into the atmosphere. The resulting low-lying marine cloud would reflect the sunlight. Daedalus has proposed a diesel fuel whose particulate emission does the same job.[5] Water vapor from the air would condense around each particle, giving a floating reflective droplet. He has also proposed reflecting the sunlight by giving commercial ships a special stable, frothy, reflective wake.[6]

Reducing Humidity

One way of looking at any problem is to see how nature solves it. In the case of reducing humidity, even this fails. Nature seems to accept humidity as simply given. Even those "air plants" alleged to need no watering, because they get water from the air, get it as the liquid droplets of mist or fog.

Now the humidity of air matters. There are about 10 grams of water in every cubic meter of air. How splendid to extract it for human use! The byproduct, dried air, would be useful too—in clothes dryers, hand driers, and so on. Many human settlements, not to mention ships at sea, would welcome a free aerial source of pure water.

Yet neither nature nor humanity can get easily at aerial water. If the humidity of the atmosphere exceeds 100%, fog or clouds form, and the excess can fall as rain. This drains in rivers to the sea, or collects as "fossil water" in underground artesian basins. Human towns cluster on rivers, not merely for the transport, but for the fresh water they collect.

We cool the body naturally—via perspiration. This relies on low humidity. A modern air conditioner reduces humidity too, but brutally and inelegantly. It cools the air so much that liquid water comes out of it, as fog or fluid. It is essentially a refrigerator for the air and uses a lot of electricity. The dehumidifier is another expensive electrical gadget that salutes our inability to control humidity. But could we reduce humidity more neatly? Daedalus has proposed several ways of doing this; perhaps his "hygroller towel" is most feasible.[7] It depended on reverse osmosis through a semipermeable membrane (fig. 15.3). It still makes sense!

Burning Waste under Water

I love this idea but have failed to imagine its chemistry. It should be chemically possible to burn any waste under water. Primitive societies just dumped unwanted stuff in pits (to the benefit of modern archaeologists). But paper-based bureaucracy and plastic packaging have made us all into copious dumpers. Some waste can be burned under water already.

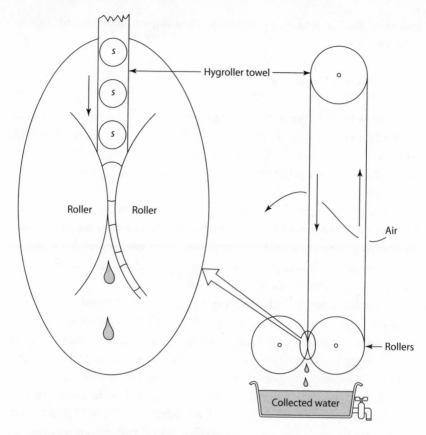

FIGURE 15.3 Daedalus's Hygroller Towel
The towel absorbs the moisture of the air and squeezes it out as water. An end-less loop of fabric is loaded with semipermeable beads (s), each containing a hygroscopic solution of calcium chloride. These beads take the humidity of the air and pass through pinch rollers, releasing the water they have captured to a receptacle below.

J. Frankham explains how the trick recovers precious-metal catalysts;[8] and some pharmaceutical companies destroy biologically dangerous wastes by burning them under water. Current techniques use high temperatures and high oxygen pressures. Biology is far cleverer. Digestion, for example, goes at body heat and normal pressure.

Such an "aqueous trash can" would complete the material cycle of our throwaway society. In a sense it would regenerate the old domestic coal fire, on which much rubbish was burned for heat. It would replace both the waste-disposal system and the sewage one. You might even distil its copious hot water to pure new domestic water. It might be even more use if it burned metals as well (tins to rust, for example). Some solid wastes, such as ceramics and glass, it could not touch. But it would certainly upstage the conventional incinerator, whose nasty smoke is widely disliked.

Material Inventions: New Stuffs and Methods

Out of the thousands possible, I have chosen four. Each seems to fill an obvious gap; together they illustrate one outcome of my playful musing on the technical scene.

1. *Filter to separate the gases in air*. Air is a mixture. A cubic meter of it contains about some 1,300 grams of air. Of that, 980 grams are nitrogen, 300 are oxygen, 17 are argon, 10 are water vapor, and 0.6 are carbon dioxide (these last two are very variable). I have already enthused about extracting the water vapor (see above). A more complete separation would be even more valuable. And it might be easy: that ancient biological invention, the gill, already does it for oxygen. Selective filtration should take things much further. We might get there soon by pure accident.

Even a rough separation would be highly useful. At present clumsy and heavy cylinders of pure oxygen are used in welding, glassblowing, and medicine. A gas containing (say) 80% of oxygen and made by membrane filtration would do just as well.

We might also concentrate the argon. This heavy gas could transform domestic heating. Hot air always rises, so domestic heating mainly warms the ceiling. Daedalus once claimed (fallaciously) that you could stop that rise by adding methyl formate (see chapter 6). But argon could do it for real! Argon has a molecular weight of 40, compared with an average of 29 for air. Hot air enriched with argon should indeed stay down. You would want to keep 20% of oxygen in the gas, to avoid suffocating cats, hamsters, and similar ground-based domestic pets. So, says my RIG, make from the air a mixture of 20% oxygen, 40% nitrogen, and 40% argon. It would have no smell or bronchial effect, but its average molecular weight would be 33.6 compared with 29 for air. So it would stay down. Only some 50 degrees Celsius of added heating would make it rise—whereas domestic radiators only heat the atmosphere by about 10 degrees Celsius.

2. *New metallurgy*. During the nineteenth century, aluminum was very expensive. Napoleon had an aluminum cutlery set to impress foreign dignitaries. Later aluminum was used militarily, for cavalry helmets. Clever Victorian chemistry made it the common industrial metal it is today. The twenty-first-century equivalent is also light and strong—titanium. The metal itself is vastly expensive, but lots of its compounds (such as titanium white paint) are in common use. The Space Shuttle was made of titanium—cost was secondary, but lightness and strength were crucial. So we need to make titanium cheap!

3. *Tramp elements*. These are harmless impurities that accumulate in recycled material until they do damage. When iron and steel are recycled as scrap, for example, the tramp element copper slowly builds up and cannot easily be removed. The archetypal tramp element is sodium (hence the salt in the sea). We use sodium compounds in soda, soap, detergents, and glass, not for the sodium but for the chemical bits that go with it. And yet it is the sodium in the water that limits our ability to recycle sewage onto the land. Some nonmetallic, easily oxidized chemical entity, such as tetramethylammonium, might be invented for soap and detergents instead.

4. *New glue*. Early colonists often impressed indigenous peoples with their clever glue; yet many modern materials are glueless. We have no glues for the meltable plastics such as PVC, polyethylene, polypropylene, and so on. A plastic object is often quite useless if a hinge or other small bit breaks off. Even superglue does not work. I'd like a good glue for thermoplastics and, while we are at it, one for metals, though I would trust good old lead-tin solder more than any glue.

Identifying People

The Nazis had a highly effective way of identifying concentration-camp victims, proof against any sort of impersonation. They tattooed a unique number on the arm of each victim. In Auschwitz, Primo Levi became Häftling 174517. Nobody wants to emulate the Nazis, but every society now needs a sure form of personal recognition. Nearly everyone fears identity theft and is concerned about unknown terrorists slipping in and wreaking havoc. At present each of us is lumbered with a huge set of numbers and passwords for machine transactions, and things can only get worse. The British government once planned to solve the problem with a highly insecure identity card. Technologists have tried (e.g., retinal scanning), but no existing technology seems foolproof. DNA is perhaps the best bet; yet getting a sample is intrusive and may recall the Nazis. Many animals already have objects called chips implanted into them to give them identity. My RIG imagines that we might each welcome a little tattoo or a subcutaneous bit of metal, carrying a number or a barcode, as a machine-readable identifier.

Getting Our Meat by Tissue Culture

At present we get meat by killing live animals. We have done it for thousands of years; predators and carrion-eaters have done it since animals began; but growing numbers of vegetarians now object to it. Killing animals is inefficient in at least two ways. First, you have to give an animal about 4 kilograms of fodder for it to gain weight by 1 kilogram. Second, every society has only a small amount of the animal that it will happily eat. In the West, for example, we eat muscle tissue. Other parts may need to be disguised, in the form of sausages, for example. Worse, we are often very cruel to our animals—think of pigs crammed into small pens and chickens into coops and geese overfed to make paté de fois gras.

Can tissue culture work instead? It has been around in the laboratory for decades. I like the idea of making meat cells on a vast scale by tissue culture, rather as we make alcoholic drinks by brewing. It would need no inefficiency or cruelty. Nobody has any idea how to do this; but it is not inherently absurd. Daedalus has proposed a sort of intermediate technology.[9]

Intellectual Inventions

Einstein felt that science itself depended on two inventions. One was the axiomatic method of reasoning, invented by the ancient Greeks and shown at its best by the geometry of Euclid. The other was the experimental method, invented by Galileo during that mighty creative ferment, the Renaissance. I would like to add a third invention, the scientific journal. This allows an argument to be pursued through space and time (previously scientists wrote books, or sent letters to each other). The earliest scientific journal I know of is the Italian *Il Nuovo Cimento*, but the *Transactions of the Royal Society* was early in the field. There are now thousands of scientific journals, most of them highly specialized. They have a style of their own (see chapter 14) with many conventions. Together these journals comprise "the scientific literature" (sometimes just "the literature"). It exists as bound volumes of past journals, held in huge libraries, but now slowly being transferred to computer storage. One convention was soon established in the journals: an experiment must be repeatable. Your worst enemy must be able to do it from your description of it and will be forced to admit that the result is what you said. As Paul Valéry wrote, "Science is the aggregate of the recipes that are always successful." Indeed, a good scientist always does a new experiment twice, not merely to discover the way the thing varies but also to reassure himself that it really happens. Without repeatability, the most accurate records are useless. (The *Annals of Irreproducible Results* is a jokey scientific journal whose title salutes this.) One damning phrase in serious scientific journals is "in our hands. . . ." It often opens a description of how an experiment or technique could not be repeated.

As a result, there is a huge penumbra of human observations that are not "scientific." Individual reports of telepathic knowledge, curious coincidences, ghosts, single observations, good and bad luck, in fact much of life, all must be dismissed as "anecdotal evidence."

Now one of the great triumphs of science has been its demolition of magic. You really don't have to say a special form of words over the microbalance; you don't have to conduct a chemical reaction during the right phase of the moon. Blessings and curses do not work. Almost all the doctrines built up before science came along have simply been shown to be wrong. (Think of all the absurd prescientific theories of what supports the Earth, for example.) The direct, rational, scientific way of doing things *works*. And yet I wonder if we are throwing the baby out with the bathwater? We need a fourth scientific invention, some way of handling pesky individual observations, and I cannot see any way of getting it.

Pushing against Our Own Wider Ignorance

We cannot prevent or cure cancer, or schizophrenia, or heart disease, or AIDS, or Alzheimer's disease, or many tropical parasitic diseases. We cannot even treat widespread minor ailments, like tooth decay or aural tinnitus or the common cold or body scarring. Sex is still dangerous, fertility hard to increase or reduce, and the sex and character of any offspring impossible to decide in advance. We still have no idea what life is about and the place it may have in the universe. Evolution theory builds on what we all know: that every living thing—including every human being—will die. Further, many human sense organs (such as our eyes, ears, and balance) decay or deceive us toward the end. Even pain is hard to control. Some biological quirk makes most major effective painkillers addictive.

We cannot even decide when somebody is effectively dead. The criteria of "brain death" are woolly and ambiguous. In 2005, after 15 years in a "persistent vegetative state," the American woman Terri Schiavo was allowed to die by the withdrawal of her feeding tube. There was a huge national outcry. That case resonates for me: I have been in such a state, but luckily came out of it. Luck was all the medics had. We still have no understanding, let alone any technical control, of human consciousness (see chapter 16). We just rely on human reproduction to go on producing conscious beings. Often Nature fails to deliver. Human reproduction is very chancy compared with that of (say) farm animals.

Yet even this human tragedy can spur mighty creativity. A famous female UK writer took up her craft after several unsuccessful attempts to have a baby. And the UK newspaper *The Times* once carried this proud announcement in its "births" column: "To Colonel and Mrs Smith. Gratefully, after seven years. A telephone."

16

A List of Silly Questions

This chapter is a tiny subset of the many things I don't know. No textbook has satisfied my curiosity. Each question may, of course, have a perfectly good answer that I happen not to know. Even so, I sense that each topic hides something to be found out and is a promising browsing region for the creative mind. Everyone should keep such a list: partly to remind you of all the thing you don't know and partly as a steady challenge to your RIG.

We all accumulate such puzzles. I advocate noting them down and wondering about them every so often. Your list will change all the time and will in any case be entirely different from mine. But here is a snatch of mine at the moment.

How Come We Get Motion Sickness?

NASA runs a plane that travels in a vertical parabola and can maintain zero gravity for perhaps 20 seconds. It is called the "vomit comet." Zero gravity induces motion sickness in many of those who volunteer for astronaut duty; the vomit comet helps to screen them out.

Now any organism can be disturbed by variations to its inertial and gravitational world. But why does this cause stomach upset in humans? The theory of motion sickness, if there is one, alleges that some "toxins" cause upsets in the inner ear, which upsets a person's balance. The body reacts by emptying the stomach to get rid of the toxins. Now I know only two such toxins, both products of civilization—alcohol and heavy water. And if humans evolved from creatures that swung in trees, we should be relatively immune to gravitational disturbances anyway. What other creatures suffer from motion sickness? Many humans, and most sailors, can get used to bodily motion. I do not understand.

Why Do Leaves Have Chemicals?

Chemicals in leaves include the alkaloids, cocaine, theobromine, caffeine, nicotine, and so on. I was told as a student that because plants cannot excrete, they dump everything they cannot use in their leaves. When they shed their leaves, they get rid of the unwanted material. Modern teachers say that leaves are designed not to be eaten. Insects are their worst enemies, so leaves tend to contain insecticides. Nicotine is the most famous, and pyrethrin may be the second most.

Yet leaves seem very bad at not being eaten. Really effective insecticides, such as DDT, are human inventions. One day, perhaps, plants will invent the deadly organophosphorus insecticides. Meanwhile, all creatures depend on eating plants or eating the creatures that eat them. Plants seem to defend themselves mainly by putting a cellulose wall around each leaf cell. This discourages humans, but many animals can digest cellulose.

And what about alkaloids? Medicine still values them. In the old days, it depended almost totally on chemicals made by plants. How come the plant spices? Did they evolve as insecticides?

How Does a Baby Spider Migrate on a Thread?

I once tried a calculation that completely failed to support my intuition about a fiber in the air and how it could drift down or rise up.[1] Yet millions of baby spiders rise in the air this way. The baby spider matters, too. A detached thread just drifts down and becomes a piece of gossamer.

Where Is the Oil Filter in the Blood?

The manner of the circulation of the blood was proposed and proved by William Harvey in 1628. Venous blood, he said, is pumped by the right side of the heart to the lungs. Modern biochemistry asserts that it dumps carbon dioxide there and picks up oxygen. Then it goes back to a second pump unit, the left side of the heart. From there, arteries and capillaries take it to all the tissues of the body. Each cell absorbs oxygen and food from a nearby capillary and dumps waste products such as carbon dioxide into it. The blood flows on, the capillaries reunite into veins, and ultimately one big vein empties that blood into the right side of the heart again.

This reminds me of the circulation of oil in machines. A crucial component of such a system is the oil filter, which removes irregular suspended matter picked up by the oil in its travels. So where is the oil filter of the body? It must have one; I cannot believe that blood never picks up anything that needs to be removed. Indeed, a clot drifting in the blood may jam in a vessel. In the brain this causes a stroke. Air passengers who stay still for many hours may get deep-vein thrombosis in their legs. Such events are rare. Most of the time, some blood-filter must mop up these nasties. What is it?

How Do Flies Walk on or Stick to a Solid Surface?

A fly seems to walk freely but takes off very promptly. I feel that the insect foot grips voluntarily and just lets go when the insect wants to take off. This argues against the usual explanation, molecular adhesion—the way solid surfaces attract each other. If the insect stuck on that way, it could not then take off. The same argument applies to wet feet, another theory of insect attachment. Furthermore, dust does not seem to cling to glass, although it is much lighter than an insect.

I have seen a fly walking on the slippery polymer PTFE, and I have even seen one standing upside-down on a "ceiling" of wet ice. Furthermore, insects are well known in the Arctic, where they must cling to ice and surfaces well below the freezing point. Indeed, I have never seen any insect sliding down a surface it was unable to grip.

The adhesion seems not to be air pressure either, though a fly under vacuum soon loses its grip. And cooling the glass seems not to freeze a fly to the spot, though it soon gets torpid. Chloroformed on a windowpane, a fly just curls up its legs up and falls off. I like the idea of killing a fly on a window instantly, say with a sudden blast of neutrons. Would the corpse continue to cling to the window?

I have also studied flies' feet through a glass prism, under a magnifying glass. The molecular contact, closer than a wavelength of light, should give a very clear "footprint." I never saw anything.

Why Do Flames Flicker?

My first guess was that their fuel supply fluctuates. Indeed, a fake electric candle used in at least one Newcastle Indian restaurant, switches the battery voltage up and down, so that it flickers. Even Faraday reckoned that a candle flickered because of the varying flow of the molten wax rising up its wick. But a detached flame driven from a constant supply (a propane canister) still flickers. I now think that, like many reactions, a flame is inherently unstable. It flickers anyway. Incidentally, there is an infrared fire detector that looks for flicker. It ignores steady sources of infrared such as lights switching on. But it might well sound the alarm in an Indian restaurant.

What Makes a Spark?

A spark can result when two solids are struck together. There are millions of solids and therefore billions of possible sparking combinations. Yet the only sparkers I know are the metals iron, titanium, and cerium, when struck by something of higher melting point (such as a hard rock). The mineral iron pyrites can also make sparks. The sparking "flint" in a lighter is an alloy of iron and cerium. Before matches were invented, fire was made using a tinderbox in which a rock such as flint struck steel. Steel is, of course, an alloy: mainly iron with some carbon. Carbon itself is a special case. Most fireworks spray out carbon sparks. A quote from the Bible says "Man is born unto trouble as the sparks fly upwards,"[2] suggesting that the writer had never seen a spark descending. I have never seen sparks discussed in any chemical textbook and do not know what is special about the sparking metals. Engineers have told me that no other metals spark on a grindstone. I have never got sparks from a carbon rod pushed against one.

Why Are Echoes So Rare?

I once burst a balloon in my sitting room. I recorded the sound on my old open-reel tape recorder and played it back slowly. The tape seemed not to hold any echoes. This contrasts strongly with Erasmus Darwin's notes on the blind Judge Fielding. After speaking a few words he said,

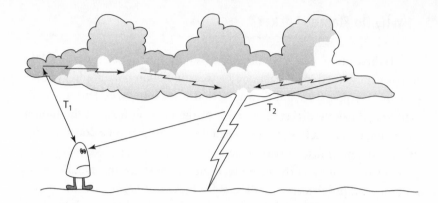

FIGURE 16.1   Rumbling Thunder
A spread of thunder noises (T_1 to T_2) following a lightning strike reaches an observer.

"This room is about 22 feet long, 18 feet wide, and 12 feet high." He was right! He must have learned to judge echoes very sensitively. I was once in a child-proof school. The architect had designed a set of concrete rooms that were highly echoey even to me. And yet clear formal echoes are rare. In a 1950s competition for the most misleading advice one could give a foreigner visiting England, Gerard Hoffnung had a brilliant entry: "Try the famous echo in the British Museum Library reading room," he said. Delightfully, it has one!

Why Does Thunder Rumble?

Thunder can rumble on for a minute or more. Lightning itself is a big atmospheric electric spark. It expands the air as it traverses it and makes a simultaneous crack of thunder. But while light is almost instantaneous, sound moves through air very slowly (3 seconds per kilometer). The noise of distant lightning can be delayed by many seconds while it travels to us. But why is it not sharp when it arrives (fig. 16.1)?

Wilson Reddish, a top ICI expert on electrical insulation, sees a thunderstorm as horizontal. The whole cloud deck goes electrically negative with respect to the Earth. It ultimately breaks down at one point, where the initial lightning strike falls. The falling strike has such a low resistance that by Ohm's Law, the cloud deck goes almost to earth potential (that is, zero) at this point. The whole cloud then shorts through it and keeps the primary stroke going. Reddish's notion is that thousands of horizontal lightning strikes then happen in the cloud. It may shroud them; they are certainly dim compared with that bright, powerful, deadly vertical strike. Yet they launch most of the thunder. It rumbles because the hearer is close to the near part of the cloud deck but farther from the far part whose noise takes much longer to reach him. NASA has reported that astronauts have seen thunderstorms from above. Some are indeed horizontally extended. I have not seen a textbook explanation.

Why Are Frozen Patterns on Cars and Windows Curved?

Many ice patterns grow from atmospheric water vapor and build up on windows, car roofs, and other flat surfaces. I have never seen a linear one (fig. 16.2). All crystals should have straight edges (they should be needles, prisms, and so on) because the molecules composing them are all identical. This is what a chemist finds in the test tube. But in free air, with water vapor deposited as ice, we get these splendid swirls. Maybe the temperature of deposition changes all the time, changing the dimensions of the deposited solid. But I do not know the answer and have never seen the effect commented on in any textbook.

How Do Plants Lift Water?

I mention this problem in chapter 2. All land plants take in water from their roots and transpire it as vapor into the air from their leaves. Even watercress needs some leaf structure above the water through which to transpire. Daedalus once proposed getting salty water into a plant by increasing the hydrostatic pressure outside it.[3] It turned out that A. Termaat had tried this seriously a little while before Daedalus had.[4] It did not work.

Why does sap rise? Is it the suction of water evaporating from high leaves? One major snag is that suction cannot make water rise more that 10 meters. Yet many trees are more than 10 meters tall!

Figure 16.2 Frost on a Window
Chemists are used to crystals with straight sides. But water vapor deposits as frost on a window, or on another cold surface, in splendid curves and swirls. Indeed, I have never seen straight frost. What is going on?

One widely accepted theory is that sap is under tension and negative pressure. This worries me. It does however explain why the liquid ducts in plants are always very narrow, typically 0.1 millimeter or less. It stops big bubbles forming in the sap—there is no room. Conversely, in animals (which are always under pressure) blood vessels can be many millimeters wide.

How Do Brain Cells Store Information?

The short answer is that nobody knows. One theory is that several neurons (brain cells) can be connected in a loop. Once started, a pulse goes around and around; the loop "reverberates" and stores a bit of information. On this theory, the brain stores data purely dynamically; the cells themselves are unaltered by new knowledge. Daedalus has exploited this theory by looking in the brain for circulation frequencies and has planned to test it by studying brain knowledge stored in frozen mammoths.[5]

A more modern motion is that each brain cell stores information on its own. Each has thousands of inputs ("synapses") coming in from other cells. Some of these activate the target cell, while others inhibit it. I imagine that each incoming pulse synthesizes a small amount of some activating or inhibiting chemical neurotransmitter. This joins the cell's stock. If the stock reaches a critical excess of activators over inhibitors, the cell fires in its turn and sends a pulse to all its own connections. Donald Hebb has proposed that the mere use of an input strengthens it, so that it synthesizes more neurotransmitter per pulse. The idea is that a pretty planless network (and the brain seems to be one) can somehow learn. If somebody devised an electronic neuron that "learns" in the same way, you could couple lots of them together at random and test the theory. You could see if the assembly could learn. At present, I am ignorant and baffled. My brain presumably knows, but is keeping quiet.

Why Are We Conscious?

Consciousness is one of those crucial things that nobody understands at all. Why should anything made of atoms be conscious? And yet I am, and I grant consciousness to all human beings and to many animals. Many books have been written about the problem;[6] I have discussed its chemistry.[7]

Now consciousness means being aware of observations, as opposed to just reacting to them. One theory of it maintains that any data-processing system can be conscious. (HAL, a computer in Kubrick and Clarke's film 2001: A Space Odyssey comes to mind.) Another claims that some special apparatus, a "soul," is needed to make an entity conscious. Yet another posits that a lesser apparatus, an "anima," suffices. A soul survives death; an anima does not. An animal is commonly taken to have only an anima. Buddhist theory makes no distinction: at death, either mental principle loses all memory of its past life. It may then enter a newborn creature and make it conscious.

Meanwhile, there is the Turing test. Alan Turing, the great computer pioneer, suggested a test in which a human judge faced two teleprinters.[8] One was connected to another teleprinter controlled by a human being; the other went to a machine. The judge could type anything he wanted on either of the teleprinters and could receive its reply. If he was unable to tell which teleprinter went to the human being and which to the machine, then you have made a machine that can think. In a sense, the judge compares the consciousnesses presented to him with an authentic sample—his own.

Daedalus has mused on consciousness.[9] He wondered how much information could be carried by a human soul. He decided that it had to carry 33 bits of information merely for a name to distinguish it from other human souls; so it might carry more.

And suppose consciousness arose at some point in evolution—as most biologists seem to assume. Then, said Daedalus, it must be represented somewhere on the genome. Creatures that are conscious should have a set of DNA genes absent from those that are not conscious. And consciousness is a remarkably unitary phenomenon. It can be abolished by a wide variety of very simple molecules—anesthetics—without affecting most other bodily systems. So it is probably coded for by one, or just a few, genes. How to identify them?

Daedalus thought of alcoholic "palimpsest." In the advanced stages of his disorder, an alcoholic has absolutely no memory of some past episode, even though at the time he did not appear drunk. So perhaps alcohol, itself a considerable anesthetic, erases not the memory, but the consciousness of its victim. It leaves a perfectly functional human robot, seemingly normal and responsive, but in fact with no internal awareness. Daedalus set DREADCO's biologists to look for the DNA of consciousness and for ways to disable it.

They created a new drug, Nothingness. A human user on Nothingness will seem entirely normal. But he will be a pure robot, talking and reacting in all the usual ways but without feeling. Behind his fluent mannerisms and animated face, inside his skull, there will be nobody at home. How will Daedalus tell if Nothingness works? He argues that no robot could ever judge, or could even have dreamed up, the Turing test. So he is inventing a "meta-Turing" test. A robotic, unconscious man will reveal the fact by being quite unable to judge a Turing test.

Perhaps the most puzzling aspect of consciousness is that we are all different. I am I, born in 1938: before that is irretrievably the past, after my death is irretrievably the future. I am aware of my own sensations and memories; I have to deduce those of every other living thing. Each individual puts his or her own data into this statement. In Buddhism, we may all go through unlimited reincarnations as a man or as an animal: "bound on the wheel" until released from this cycle to enter "Nirvana." I ponder that any memories of previous lives are suppressed into the unconscious so that the new creature concentrates on being whatever it is.

Science says nothing on any theory of consciousness. Perhaps we should look for mistakes. Some human might be born without a soul, giving a natural zombie detectable by Daedalus's meta-Turing test. Another might have two souls, giving perhaps a double personality. He might even survive a normally lethal fate, but with a single personality—one soul having died and escaped. Meanwhile, H. G. Wells has written an ominous story about the whole conundrum. His Mr. Elvesham could transfer his whole identity to a new young body when age began to threaten. And Woody Allen has turned it into a joke. Asked to name his greatest regret in life, he replied, "Not being someone else!"

Why Does the Eye Decline?

Many of us need glasses even in youth, and we all need them in old age. Why? We get about 80% of our knowledge from our eyes; yet they seem unable to repair imperfections and decline markedly throughout life. Does this happen to animal eyes as well?

How Poisonous Is Arsenic?

Some inhabitants of Bangladesh get their drinking water from tube wells. It has 0.05 parts per million of arsenic in it. This makes them ill. Curiously, far more contaminated European water has been claimed to be medicinal. The spring water of Court St. Etienne in Belgium may have 5 parts per million of arsenic. In the nineteenth-century Fowler's solution, with 10,000 parts per million of arsenic, was used in medicine. There was even a claim that patients "got used" to high doses of arsenic. The peasants of Styria (in Austria) were said in old textbooks to consume as much as 0.3 grams of arsenious oxide in a single dose to improve their breathing when climbing mountains.

These claims puzzle me. One of my great triumphs was the discovery of arsenic in Napoleon's wallpaper. After the battle of Waterloo, Napoleon was imprisoned on the island of St. Helena in the South Atlantic. He lived in Longwood House from 1815 and died there in 1821. In a BBC radio talk produced by Martin Goldman, I asked if any listener knew anything about Napoleon's wallpaper in Longwood House. Amazingly, I had a response! Shirley Bradley in Norfolk had an actual sample of his wallpaper—in an old scrapbook. I borrowed it, and with Ken Ledingham of Glasgow University, we analyzed it non-destructively for arsenic, by x-ray spectroscopy. We found 0.12 grams per square meter in the sample.[10] In those days the poison copper arsenite was a popular green pigment for wallpaper. Much later it was found that, if the room got damp and moldy, mold on the wallpaper could liberate a vapor, arsenic trimethyl, into the room. Breathed by the occupants, it could make them ill. Several samples of Napoleon's hair have been retained as keepsakes, and arsenic has been found in them.

Ken and I wrote a serious scientific paper on the topic; I also wrote a less formal and more human one.[11] The story even got on TV, as part of Hendryk Ball's *The Human Element* series, transmitted in 1992. One outcome was that I was offered more samples of Napoleonic wallpaper. Some samples just copied my original; but one was most interesting. In the days before the Suez Canal, many ships passing from India to Britain called in at St. Helena for water and supplies. E. D. H. Johnson of the Channel Islands had a sample of Napoleonic wallpaper. It had been taken in July 1824 from Longwood House by Captain Mitford of *The Ganges*. Captain Mitford secured a sample of wallpaper—in that damp climate the stuff was easily detached—and wrote a little note describing his adventure. His bit of wallpaper included a chunk of green dado, which I hoped would be almost pure copper arsenite and would underline my thesis.

The story was supported by an expert on *Lloyd's Shipping Register*, who confirmed for me that Captain Mitford was indeed the master of *The Ganges* and that it sailed from Bombay to Liverpool in 1824, visiting St. Helena on the way. Unfortunately, when I took the new wallpaper to Ken Ledingham (he was with Strathclyde University by then), we could not find any arsenic.

It might be there, of course—x-ray spectroscopy can be troublesome—but not in large amounts. So our original conclusion still stands. Napoleon would have died anyway from a stomach ulcer. Local officials might have fiddled the autopsy to exculpate wicked poisoners, of course. But I believe the man who should not have been there—Surgeon Henry, whose ship was passing St. Helena at the time. Arsenic in Napoleon's wallpaper might have speeded his death a little, but not much. Historians who argue that he was deliberately killed by the British (or perhaps by the French, who did not want him back) will have to depend on other documentary evidence.

These events have made me think hard about arsenic but have not given me any believable numbers. Arsenic (whose oxide was once called "inheritance powder") was a much-used poison until the sensitive Marsh Test made it easy to detect. Nineteenth-century medicine countered many diseases with sublethal doses of lethal substances, such as arsenic. I am glad that filters containing iron can greatly reduce the hazards of Bangladeshi water, but I do not understand what amount of arsenic gives rise to what physical symptom.

Why Does the Sun Rotate More Slowly at the Poles?

Everything rotates. To study solar rotation, you time sunspots. They show that the sun rotates once in 25 days at its equator but once every 34 days near the poles. The "gas giant" planets show the effect too. Jupiter rotates once in 9.842 hours at its equator, but once in 9.928 hours near the poles. Saturn rotates once in 10.23 hours at its equator, but once in 10.63 hours near the poles (fig. 16.3, panel a).

All these objects are fluid. They can easily tolerate different latitudinal speeds. Nonetheless, they have all been spinning for billions of years and should long ago have reached a uniform rotation rate. Some sort of internal "engine" must drive this rotational difference. I cannot guess what it is, nor whether it is at work in all spinning fluid objects.

What Defines the Shape and Sharpness of Stars?

The sun is a ball of gas. I would expect it to be vague and woolly, with no sharp edge. Yet the surface is wonderfully well defined, sunspots and all. How come?

We are used to the idea that stars are spherical, like the sun. Yet to even the best telescope, the biggest star is a mere unresolved point of light. As far as its image goes, it could be any shape—sausage-shaped or even cubic! So I was very pleased when a paper in *Nature* described peanut-shaped and toroidal water drops, created by spinning.[12] My RIG immediately imagined similar stars. Toroidal black holes and neutron stars have been considered by theorists, but ordinary stars may not have been (fig. 16.3, panels b, c, and d).

What Flows Can Be Reversed?

In effect, Ohm's Law says that electricity is linear. If you double or reverse the voltage, the current doubles or reverses too. Physics seems to assume that other flows, such as those of heat or light, are also linear—so that light sent through any optical system will retrace its path exactly if sent back. But electricity is not linear: ask the makers of diodes and rectifiers! So maybe other flows can be nonlinear as well. I like the idea of a brick that admits heat to warm the house when the outside air is warm but retains it when the outside air gets cooler. Similarly, a rectifying window would let sunlight in easily but not out so easily. Are such things possible?

Is the Reaction of Sodium and Water a Fuel-Coolant Explosion?

When a hot liquid is poured into a cold one, heat flows. If the cold liquid boils, you can get an explosion. Thus molten iron should never be poured into a wet crucible. The resulting steam explosion may throw molten metal dangerously about. And the Krakatoa volcanic eruption—perhaps the biggest explosion ever recorded—may have been due to several cubic miles of molten lava entering the sea.

Now it is an entertaining chemical trick to put a little bit of sodium or potassium metal, say about the size of a pea, into water. The stuff fizzes, rushes energetically about on the water surface, and can even inflame. But a big bit in water may explode dangerously. Sodium melts at 98°C, and potassium at 64°C, so both melt below the boiling point of water. My

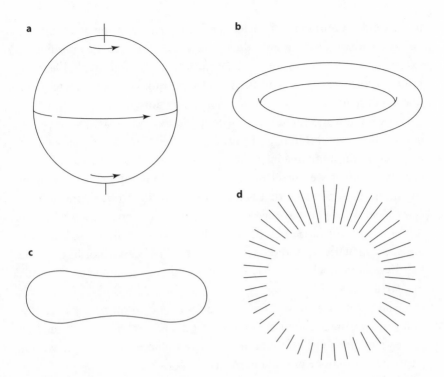

FIGURE 16.3   Shapes of Stars

Our own sun (a) rotates faster at the equator than at the poles. So do the "gas giant" planets Jupiter and Saturn. Toroidal (b) and "peanut-shaped"(c) water droplets have been made. I can imagine rotating stars being deformed like this, but astronomers do not know of any. Stars seem to be sharp, like the sun. But if a star is just a big ball of gas, it might be "fuzzy" (d).

guess is that the reaction melts the metal, and the reaction energy stirs the melt into the water and makes a fuel-coolant explosion. I can imagine ways of trying it!

How Real and Constant Are Radioactive Half-Lives?

Currently (2009), the British government wants more nuclear power stations. We can thus go on wasting electricity without releasing the carbon dioxide of burning. Yet nuclear power is vastly unpopular, essentially because we cannot clean up radioactive waste. Reactors are enormously expensive to "decommission," even if the job can be done at all. I fear the future British landscape will be dotted with radioactive concrete hulks that we dare not approach. Similar hulks may dot Japan.

The human argument is compelling, too. Nuclear waste is a menace, to us and to our children, and to their children. The one thing we have of value is the human blueprint, our DNA genome. No matter what ghastly mistakes we ourselves make, our children will grow up human. Anything that threatens that is a Faustian bargain to be avoided at all costs. And once you have made something radioactive, say by putting it in a reactor, or indeed by building that reactor in the first place, you are saddled with it. A radioactive object will decay according to the half-lives of its nuclei and may be biologically dangerous for thousands of years.

And yet nuclear power has a big appeal. A coal-burning power station can consume tens of thousands of tons of coal a week. A nuclear power station might consume less than a millionth of that, in weight terms, as uranium or plutonium for its reactor.

Sadly, its waste product is highly radioactive and "reprocessing" can recycle only some of it. The rest just decays, slowly. All nuclear technology is dominated by that slow decay. In one half-life, half the stuff decays; in the next, half of the remainder . . ., and so on. The process never really ends.

So what to do? One rather technomanic answer is to fire our radioactive waste away from the Earth, in big rockets. Some serious commentators have suggested this, and so has Daedalus—in 1971 he proposed a nuclear rocket that could lift itself.[13] Well, rockets of any kind are very expensive and not very reliable. The technique has been tried, sort-of, for spacecraft intended never to return to Earth. Thus the spacecraft *Galileo* and *Cassini*, launched to explore the outer planets, were powered by ferocious plutonium-containing thermonuclear generators.

So we come back to the question posed in this topic. How unalterable are radioactive half-lives? I know only one reference that implicitly questions half-lives, and yet my unconscious mind would love to speed them up.[14] Given an unstable nucleus, tickle it and make it decay at once!

You would (a) get a lot of useful energy and (b) be left with an inert, non-radioactive object.

Speeding up a radioactive half-life is bold but not obviously absurd. The atom bomb is, after all, a mere speed-up of the nuclear-fission mode of decay of uranium or plutonium. My RIG dreams of hitting a nucleus with a gamma ray whose wavelength is about the same size as that nucleus. The resulting resonance might drive the nucleus into decay. A nucleus is perhaps 10^{-15} meters across; the shortest gamma rays we can make are several times longer, at maybe 10^{-14} meters; so we have a little way to go yet. And it would be hard to aim a gamma ray at a nucleus, and hard to tune the collision properly.

And yet I dream of the project. The alternative seems just to keep radioactive waste safe while it decays. Yet what containment will stay safe for 10,000 years? An American plan is to put waste in a deep geological shaft. An Australian one advocates a "synrock" to imitate the geological rock that has held radioactive materials for millions of years. The radioactive waste of Sellafield in the United Kingdom was intended to be poured into the Irish Sea, until the Welsh and Irish got spooked and objected. I have heard of a plan to dump radioactive waste into a jungle, keeping human predators away while speeding evolution locally. The surroundings of Chernobyl do not encourage this idea.

What Determines the Physical Constants?

I have a verse on this matter (see chapter 14). Freeman Dyson observed that the nuclear attraction forces are very cunningly balanced.[15] The strong force holds each atomic nucleus together. Yet it is just too weak to allow two protons to cling together stably into a "diproton." If the diproton existed, ordinary protonic hydrogen would be very rare. So would stars like the sun, burning that hydrogen to helium. That steady burning has allowed the sun to maintain a stable surface temperature for billions of years, perhaps giving one of its planets a chance to evolve life.

Dyson goes on to present other odd features of the known universe. Thus interstellar hydrogen has a low density, requiring stars on average to be about 20 million million miles apart. If they were much closer than this, another star would approach the sun often enough to destroy any life evolving on its planets. Such arguments support the anthropic principle, by which the physical constants were chosen to encourage life.

What Goes on at a Surface?

In a crystal such as sodium chloride, each Na^+ ion is surrounded by six Cl^- ions, and vice-versa. Everything balances. At the surface of the crystal, this balance is upset. Somehow the crystal surface has to reach some chemical equilibrium with the air around it. What happens? I do not know. Indeed, Daedalus thought of the hollow carbon molecule (whose later synthesis, as buckminsterfullerene, gained the Nobel Prize in chemistry for those who achieved it) at least in part by wondering about those who split diamonds.[16] Such master craftsmen can cleave a diamond into two parts, thus suddenly creating two added diamond surfaces. How does the new carbon surface react with the air? It must be chemically unstable when it is first made. Some substances can supercool strongly—the ones I have played with are sodium thiosulfate and sodium acetate. The supercooled liquid sometimes crystallizes suddenly, and I have wondered whether a bit of new surface can set it off. I have tried a few simple experiments, but so far nothing has worked.

How Does the Double Helix Work?

I criticize the conventional theory of heredity in chapter 7. It holds that, every time a cell divides, its DNA unwinds and duplicates, thus giving a copy to both "daughter" cells. This is neat and clever. But the numbers are vast! The bacterium *E. coli*, for example, has a DNA molecule containing about 4.5 million base pairs. The molecule is about 1.7 millimeters long. To fit into a cell nucleus, such a double-helix DNA molecule must itself be wound into a bigger helix, and this into another (fig. 16.4)! Bigger organisms than *E. coli* have even longer molecules of DNA. I cannot imagine that so long a molecule could unwind completely without getting impossibly tangled up. And if it could, the resulting single strands could not duplicate and wind up again without equally appalling molecular mayhem. Yet this has to happen, flawlessly, every time a cell divides, which may be every 20 minutes or so. Maybe

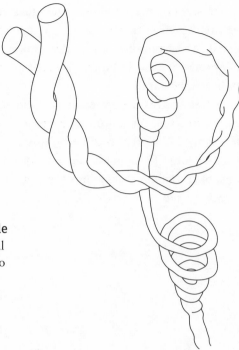

FIGURE 16.4   A DNA Molecule
To fit into a cell nucleus, a helical
molecule of DNA must wind into
another helix, then into another
and another. To replicate at each
cell division, it has to unwind
completely. Something odd must
be going on.

biochemists can imagine a mechanism, but I cannot. (A dimensional
solution is in chapter 7.)

How Do People Assess Each Other?

The usual answer is "intuition." I note it in chapter 3 but do not
understand it. I suspect that body language in any form depends on the
pattern recognition of the microsignals of human behavior, and most of
us learned to read it as children. The Royal Navy admiral Joe Bennett
once accepted into his ship the navy personnel from some maritime
disaster. The rescued men were clad only in underwear or were even
naked; yet with no uniforms to guide him, Bennett directed the officers
aft and the ratings forward. Similarly, during World War II, Eric Berne
"sized up" recruits or inductees into the U.S. armed forces.[17] All were
dressed alike, in a maroon bathrobe and cloth slippers. Berne took only

a few seconds per man and often guessed their occupations correctly before they even spoke. Socially, we gallantly pretend that everybody is equal. We seem not to notice this ranking, nor do we discuss it. "Pecking order" is clear in human beings but was first described in chickens (whence the name).

Dress makes the options much wider. At Yorkshire Television Ltd., I used to baffle the unions by putting on my white coat. Nobody then knew my authority; I could even turn a switch without the electricians walking out. Those who understand and manipulate our agreed but unspoken rankings—psychopaths and con men come to my mind—are rightly feared.

17

A Short Guide to Being Creative

*P*eople who would like to increase their own creativity may welcome a condensed account of my advice. So here's a brief summary.

General

Most of the human mind is outside our awareness. The bit we are conscious of, the Observer-Reasoner, is at the top; below it is the Censor, of which we are largely unconscious and that guards us from the totally unconscious structure below. (Figure 1.1 shows my model of this arrangement.) The stuff at the bottom includes the Random-Ideas Generator, or RIG. We all have one. Indeed, when we tackle a problem by having ideas, we are asking the RIG to send those ideas upstairs to be judged or tested. My sad guess is that 80% of RIG ideas, even those that get past the Censor, are wrong. They fail that test. The RIG is not an intellectual entity; it is emotional and plays with the material it acquires. Yet it is the basis of human creativity.

Where does the RIG get its material? The Observer-Reasoner, supplied by the senses, sends information downstairs to it, past the filtering Censor. The RIG plays with it, making odd combinations, and sends up ideas: often on demand, and sometimes after a delay, but sometimes spontaneously. Among lots of failures, it may push up a success, the Censor may let it through, and you may recognize it and take it further. This happens in all of us; at best, my advice may help to lubricate the process.

Your Large-Scale Environment

This includes your position in society and your employment. If you are a student, or occupied with routine work, or self-employed domestically, it may be relatively simple. Once you have met current demands, you are free to think and do what you like, with whatever resources may be to hand.

But suppose you do research or development? A business will probably have given you a mandate or a set of tasks. You may want to be creative about them or also to work on some project of your own. A university typically imposes a well-administered teaching load of lectures, seminars, marking, and practical classes on its staff. It may pay much less attention to your "own work," that is, your private research. Yet any creative individual must, I feel, be subtly exploitative of the administration. I once worked as a company researcher for ICI Ltd. I intuited an unwritten company rule that if money was being spent and paper was being generated, then progress was being made. So I made sure that my company tasks did both of these. My private ventures (e.g., on bicycle stability, see chapter 5) may have exploited company facilities but were less obtrusive. Early on, I established a reputation as an "odd bird." My hope was that the administration would not be alarmed if I later showed strange behavior. The strategy probably worked. I also avoided predictability and routine. I feel that the creative process can be usefully stirred up by any change—journeys, lectures, new places, new experiences, chats with strangers or "opposite numbers" in other organizations—in fact any sort of play with the opportunities that come along. Even a simple journey can spark a creative idea.

Your Small-Scale Environment

This is the world inside in your own head. We are all different, but I reckon that any creative person needs to amass a vast amount of information downstairs for the RIG to play with. So it always helps to be curious and inquisitive. I maintain a whole database of anything that appeals to me and comes my way (I describe it in chapter 14). All my life I have accumulated information of any kind on sheer whim, without imposing any filter or plan or pattern. Every so often, like junk, a bit comes in handy.

The RIG is primarily emotional. It seizes notions and ideas for their appeal rather than for their sense. The ideas it pushes up, even those that get past the Censor, are usually wrong. Most thinkers discard these wrong notions; but I reckon you should go along with them. Even if 80% of ideas fail to work, an absurd idea can sometimes be the forerunner of a bold and workable new notion.

A related strategy is "thinking with your hands": trying odd or silly experiments and being curious about small results. Again, this sends information down to the RIG. It also develops your "physical intuition," an important aspect of being a good experimenter. I feel that a creative should always be a "noticer." Let simple observations bother you—like the weak mirror image in a window or the pattern smeared out by a spinning rotor. Ask yourself silly questions. (For example, how can a powder heap up on a spoon or spatula? We all exploit that fact all the time. Yet why should any powder form a heap?) Most of the time the answer—if you find it—will not spark any useful thought. Every creative has to live with lots and lots of failure. But even one success is worth having! Furthermore, creativity can be very time-consuming. It can take months or years even to notice a problem. At the very least, I feel you should muse on a subject overnight. That gives the unconscious mind a chance to mull it over while you sleep.

Social activity in general is probably counterproductive in the creative process. Solitude often augments creativity. You should try to spend at least some time on your own. Then you can act and think as you like. I particularly advocate reverie or day-dreaming—not actually being asleep but just flopping and relaxing. Maybe the Censor is dozing too, and the RIG can push up some new idea. You may be at your most creative in bed in the morning, or while taking a bath, or while explaining something to a good listener—whom I call a "matching impedance." You need to discover the circumstances that encourage your RIG and spend a lot of time that way.

And try talking to your Censor! You want to weaken it. You should ask it to let down information, such as casual observations and odd remarks, so that the RIG can play with them. You also want it to let RIG ideas up. Even if those ideas look daft to it, you want to know about them. I have talked to my own Censor, always without response and maybe without useful result. I fear that nobody can change their Censor's strategy much in this way—but every little helps.

In particular, you want to note and encourage humor and jokes. Both going down and coming up, they stimulate creativity. Any joke is usually lost as you develop a new idea but sometimes it grows in the process. Thus my own publications on Napoleon's poisonous wallpaper were enhanced by the sheer humor of a man who was once the most powerful in the world being threatened by an item of decor. The RIG being emotional, it is always glad to get jokes from above, and some of the notions it passes up are jokey. After all, like any novelty, a joke shows something in a new light. It pays to hang on to it, to take it seriously and not to reject it just for being silly. I reckon I gained from the weekly Daedalus funny column I ran in *New Scientist* and later in *Nature* and in the *Guardian* newspaper. I always needed scientific jokes. I valued them, and occasionally, despite my best efforts, they came true on me.

Conversely, we all get depressed at times. In my view, depression shows that the unconscious mind and the RIG are being active. Depression can be very discouraging, and many of us assault it instantly with antidepressant drugs. But I reckon it is sometimes worth sticking with; it has a positive side. You may later get new ideas.

And always carry a pencil and a piece of paper. Mainly this is a waste of paper; but an idea can strike at any time, and you may be glad you made a note! Another use of paper is to maintain a list of silly questions and things you don't know. Such a list (chapter 16 is an example) helps the RIG to keep musing.

Once you have a new and worthwhile idea, it may take months or even years of hard work by yourself and others, to try it out in practice. Yet that trial is the only real test!

Notes

Chapter 1. A Theory of Creativity

1. David E. H. Jones, *American Scientist* 90 (Sept.–Oct. 2002): 454.
2. R. Trivers, *Annals of the New York Academy of Science* 907 (2000): 114.
3. Wendy Cope, "A Policeman's Lot," in *Imitations of Immortality*, ed. E. O. Parrott (Penguin, 1987), 140.
4. Patricia Garfield, *Creative Dreaming* (Simon and Schuster, 1995), 35.
5. R. A. Brown and R. G. Luckcock, *Journal of Chemical Education* (Nov. 1978): 694.
6. Arthur Koestler, *The Act of Creation* (Hutchinson, 1964).
7. Arthur Koestler, *The Act of Creation* (Hutchinson, 1964), 33.
8. Peter Medawar, *Pluto's Republic* (Oxford University Press, 1987), 252.
9. W. Platt and R. A. Baker, *Journal of Chemical Education* (Oct. 1931): 1969.
10. The Holy Bible, the Acts of the Apostles, 9:3–9.
11. H. Black, *IEEE Spectrum* (Dec. 1977): 54.
12. Freeman Dyson, *Disturbing the Universe* (Harper Colophon, 1979), 47.

Chapter 2. The Creative Environment

1. I. J. Good, ed., *The Scientist Speculates* (Heinemann, 1962), 212.
2. J. E. Gordon, *Structures* (Pelican, 1978), 376.

Chapter 3. Thoughts on the Random-Ideas Generator

1. Robert Townsend, *Up the Organization* (Alfred A. Knopf, 1970), 109.
2. Bertrand Russell, *Portraits from Memory* (George Allen & Unwin, 1956), 195.

3. David E. H. Jones (Daedalus), *New Scientist* 41(1969): 308; David E. H. Jones, *The Inventions of Daedalus: A Compendium of Plausible Schemes* (W. H. Freeman, 1982), 26.

4. David E. H. Jones, *Nature* 390 (13 Nov. 1997): 126.

5. Stephen Bragg, *New Scientist* (1 Aug. 1974): 258.

6. David E. H. Jones, *Nature* 343 (11 Jan. 1990): 122.

7. Rudyard Kipling, *Something of Myself*, ed. Thomas Pinney (Cambridge University Press, 1990), 122.

8. H. Ellis, *A Study of British Genius* (Hurst & Blackett, 1904), 191.

9. J. Maddox, *Nature* (4 Oct. 1984): 399.

10. Alan Sokal and Jean Bricmont, *Fashionable Nonsense: Postmodern Intellectuals' Abuse of Science* (Picador, 1998).

11. www.statistics.gov.uk/StatBase/xsdataset.asp?More=Y&vlnk=6114&All=Y&B2.x=618B2.y=1L

12. B. B. Mandelbrot, *The Fractal Geometry of Nature* (W. H. Freeman, 1982), 392.

13. Kay Redfield Jamison, *Touched with Fire: Manic-Depressive Illness and the Artistic Temperament* (Free Press, 1993).

14. Stevie Smith, "My Muse," in *Selected Poems* (Longman and Green, 1962), 120.

15. David E. H. Jones (Daedalus), *New Scientist* 39 (19 Sept. 1968): 615; David E. H. Jones, *The Further Inventions of Daedalus: A Compendium of Plausible Schemes* (Oxford University Press, 1999), 100.

16. V. Weisskopf, in *More Random Walks in Science*, ed. R. L. Weber, The Institute of Physics, 1982, 97, and *American Journal of Physics* 45(1977): 422.

Chapter 4. Intuition and Odd Notions

1. Konrad Lorenz, *King Solomon's Ring* (The Reprint Society, 1953), 104.

2. Eric Berne, *A Layman's Guide to Psychiatry and Psychoanalysis* (Penguin, 1968), 383.

3. R. Townsend, *Up the Organization* (Alfred A. Knopf, 1970), 64.

4. David E. H. Jones, *Nature* 370 (4 Aug. 1994): 332.

5. T. G. Goertzel and A. M. W. Hansen, *Cradles of Eminence*, 2nd ed. (Great Potential Press, 2004).

6. David E. H. Jones, *Nature* 362 (8 April 1993): 502 and 362 (15 April 1993): 592.

7. David E. H. Jones, *Nature* 333 (30 June 1988): 806.

8. K. B. Williamson, *Nature* 162 (13 Nov. 1948): 786.

Chapter 5. Creativity in Scientific Papers

1. David E. H. Jones and U. Walter, "The Silicate Garden Reaction in Microgravity," *Journal of Colloid and Interface Science* 203 (1998): 286.
2. David E. H. Jones, "Gardening in Space," *American Scientist* 90 (Sept.–Oct. 2002): 454.
3. H. Schäfer and U. Flörke, *Zeitschrift Anorganische Allgemeine Chemie* 462 (1980): 173.
4. David E. H. Jones and J. L. Wood, *Journal of the Chemical Society (A)*, Part I (1966): 1448 and Part II (1967): 1140.
5. David E. H. Jones and J. L. Wood, *Journal of the Chemical Society (A)*, Part III (1971): 3132 and Part IV (1971): 3135
6. M. Dalibart et al., *Inorganic Chemistry* 21(1982): 1040; G. V. Tsintsadze et al., *Dopovidi Akademii Nauka Ukraine*, RSR Series B 31 (1969): 130.
7. David E. H. Jones (Daedalus), *New Scientist* 29 (24 March 1966): 782.
8. David E. H. Jones, *Physics Today* (April 1970): 34.
9. David E. H. Jones, *Physics Today* (Sept. 2006): 51.
10. F. R. Whitt and D. G. Wilson, *Bicycling Science* (MIT Press, 1985), chapter 9.

Chapter 6. Heat and Gravity

1. David E. H. Jones (Daedalus), *New Scientist* 84 (8 Nov. 1979): 496.
2. David E. H. Jones, *Nature* 346 (5 July 1990): 20; David E. H. Jones, *The Further Inventions of Daedalus: A Compendium of Plausible Schemes* (Oxford University Press, 1999), 190.

Chapter 7. Astronomical Musings

1. David E. H. Jones, *Nature* 395 (1 Oct. 1998): 445 and *Nature* 395 (8 Oct. 1998): 550.
2. H. Schmidt, *New Scientist* 44 (16 Oct. 1969): 114 and 50 (24 June 1971): 757.
3. R. Rowan, *The Intuitive Manager* (Berkley Books, 1991), 141; E. D. Dean and J. Mihalasky, *Executive ESP* (Berkley Books, 1974).
4. *The Economist* (19 Feb. 2005): 86.
5. J. Giles, *Nature* (30 Sept. 2004): 494; *The Economist* (8 March 2008): 102.
6. S. Weiner et al., *Nature* 287 (1980): 820.
7. T. de Torres, J. E. Ortiz et al., *Archaeometry* 44, no. 3 (2002): 417.

8. I. W. Roxburgh, *New Scientist* 63 (26 Sept. 1974): 828.

9. Bertrand Russell, *History of Western Philosophy* (George Allen and Unwin, 1957), 480.

10. Edwin A. Abbott, *Flatland*. Original publisher Seely & Co., 1884. The Dover "Thrift" edition of 1992 has Abbott's charming original illustrations.

11. David E. H. Jones (Daedalus), *New Scientist* 25 (14 Jan. 1965): 89; David E. H. Jones, *Nature* 396 (3 Dec. 1998): 417.

12. W. Fowler, E. Margaret Burbidge, G. Burbidge, and F. Hoyle, *Science* 124(1956): 611 and *Review of Modern Physics* 29 (Oct. 1957): 547.

13. David E. H. Jones, *Nature* 356 (26 March 1992): 292; David E. H. Jones, *The Further Inventions of Daedalus: A Compendium of Plausible Schemes* (Oxford University Press, 1999), 108.

14. David E. H. Jones (Daedalus), *New Scientist* 32 (3 Nov. 1966): 245; David E. H. Jones, *The Inventions of Daedalus: A Compendium of Plausible Schemes* (W. H. Freeman, 1982), 118.

15. David E. H. Jones (Daedalus), *New Scientist* 72(1976): 72.

16. David E. H. Jones, *Nature* 334 (28 July 1988): 300.

Chapter 8. Rotating Things

1. H. K. Moffat, *Journal de Mechanique* 16, no. 5 (1977): 651.

Chapter 9. Explosions and Fuses

1. S. Roman, *Chemical Technology* (July 1990): 386.

Chapter 10. Tricks with Optics

1. David E. H. Jones, *Nature* 414 (13 Dec. 2001): 706.
2. David E. H. Jones (Daedalus), *New Scientist* 27 (30 Sept. 1965): 843.
3. M. Chown, *New Scientist* (5 Sept. 1985): 19.

Chapter 12. Physical Phenomena I Have Noticed

1. David E. H. Jones (Daedalus), *New Scientist* 72 (9 Dec. 1976): 632.
2. David E. H. Jones, *Nature* 334 (25 Aug. 1988): 654.
3. David E. H. Jones (Daedalus), *New Scientist* 58 (3 May 1972): 320.

Chapter 13. Odd Notions I Have Played With

1. J. D. Horwitz et al., *American Journal of Cardiology* 29(1972): 149.

2. J. B. S. Haldane, *The Inequality of Man* (Pelican Books, 1937). His article "The Story of My Health" begins on page 228.

3. W. B. Bean, *Archives of Internal Medicine* 134(1974): 497 and 140: 33.

4. H. A. C. MacKay, *Chemistry and Industry* (28 Nov. 1964): 1978.

5. David E. H. Jones, *Nature* (19 July 2001): 292.

Chapter 14. Literary Information

1. J. E. Gordon, *The New Science of Strong Materials*, Penguin, 1991, 141.

2. David E. H. Jones, *Omni* 2(12 1980): 130. From S. Coren and C. Porac, *Science* 198 (Nov. 1977): 631.

3. Freeman Dyson, *Innovations in Physics*, quoted in *Dictionary of Scientific Quotations*, A. L. Mackay (Adam Hilger, 1991), 78.

4. S. Marinov, *New Scientist* 116 (18 Dec. 1986), 48; S. Marinov, *Nature* 380 (28 March 1996), reader-enquiry card, 161.

5. W. M. Honig, *The Sciences* (May–June 1984): 24.

6. G. H. Keswani, *Speculations in Science and Technology* 9, no. 4 (1986): 243.

7. David E. H. Jones, *Nature* 376(27 Jan. 1994): 324; *Further Inventions of Daedalus: A Compendium of Plausible Schemes* (Oxford University Press, 1999) 172.

8. D. Underwood, *The Mathematical Intelligencer* 5, no. 1(1983): 20.

9. U. Dudley, *Mathematical Cranks* (Spectrum, 1992).

10. J. Duncan and S. G. Starling, *A Textbook of Physics* (Macmillan, 1920), 491.

Chapter 15. Inventions We Need But Don't Have

1. R. A. Herendeen, *Science* 205(3 Aug. 1979): 451.

2. David E. H. Jones, *Nature* 346(5 July 1990): 20; David E. H. Jones, *The Further Inventions of Daedalus: A Compendium of Plausible Schemes* (Oxford University Press, 1999), 190.

3. *The Economist* (Technology Quarterly) (4 Sept. 2010): 14.

4. David E. H. Jones (Daedalus), *New Scientist* 58 (3 May 1973): 320.

5. David E. H. Jones, *Nature* 401 (28 Oct. 1999): 871.

6. David E. H. Jones, *Nature* 348 (29 Nov. 1990): 396.

7. David E. H. Jones (Daedalus), *New Scientist* 105 (28 March 1985): 88 and 106 (4 April 1985): 68.

8. J. Frankham, *Chemistry and Industry* (1 March 2004): 18.

9. David E. H. Jones (Daedalus), *New Scientist* 25 (18 Feb. 1965): 436 and

33 (16 March 1967): 550; and David E. H. Jones, *The Inventions of Daedalus: A Compendium of Plausible Schemes* (W. H. Freeman, 1982), 24.

Chapter 16. A List of Silly Questions

1. David E. H. Jones (Daedalus), *New Scientist* 67 (1975): 192; David E. H. Jones, *The Inventions of Daedalus: A Compendium of Plausible Schemes* (W. H. Freeman, 1982), 166.

2. The Holy Bible, Job, 5:7.

3. David E. H. Jones (Daedalus), *New Scientist* 107 (25 July 1985): 88.

4. A. Termaat et al., *Plant Physiology* 77 (April 1985): 869.

5. David E. H. Jones (Daedalus), *New Scientist* 61 (17 Jan. 1974): 176 and 61 (24 Jan. 1974): 240; David E. H. Jones (Daedalus), *New Scientist* 40 (14 Nov. 1968): 394 and 40 (21 Nov. 1968): 446; and Jones, *Inventions of Daedalus*, 44.

6. C. Blakemore and S. Greenfield S (eds.), *Mindwaves* (Blackwell, 1987).

7. David E. H. Jones, *Chemistry World* 6 (3 March 2008): 96.

8. Alan Turing, e.g., page 261 in Blakemore and Greenfield, *Mindwaves*.

9. David E. H. Jones (Daedalus), *New Scientist* 76 (17 Nov. 1977): 464 and 76 (24 Nov. 1977): 544; and Jones, *The Inventions of Daedalus*, 50; David E. H. Jones, *Nature* 403 (20 Jan. 2000): 263 and 403 (27 Jan. 2000): 380.

10. David E. H. Jones and Kenneth W. D. Ledingham, *Nature* (14 Oct. 1982): 626.

11. David E. H. Jones, *New Scientist* (14 Oct. 1982): 101.

12. P. Aussillous and L. Mahadevan, *Nature* 411 (21 June 2001): 924.

13. David E. H. Jones (Daedalus), *New Scientist* 46 (7 Jan. 1971): 387; and Jones, *Inventions of Daedalus*, 92.

14. J. L. Anderson and G. W. Spangler, *Journal of Physical Chemistry* 77 (20 Dec. 1973): 3114.

15. Freeman Dyson, *Disturbing the Universe* (Harper and Row, 1979), 250.

16. David E. H. Jones (Daedalus), *New Scientist* 32 (3 Nov. 1966): 245; and Jones, *Inventions of Daedalus*, 118.

17. Eric Berne, *A Layman's Guide to Psychiatry and Psychoanalysis* (Penguin, 1971), 387.

Index

Abbott, Edwin, 120
absorption, 85, 87
acetonitrile, 93, 94, 154
acetylene, 136, 142, 143, 145
acetylide, 122
adhesives, 86, 89, 106, 149, 159, 222
advertisements, 45, 180, 193, 202, 206
air: actuating, 85, 87, 142, 176–78, 186;
    balloons and lift, 101–2, 105, 107–8;
    chemistry of, 91, 92, 137, 215, 231,
    242; cooling, 131, 133, 167, 218;
    deforming film, 158, 159; flame and
    explosion, 142–44, 145, 148; gases
    in, 21, 59, 219–21; heating, 100–102,
    139, 164, 221; pressure of, 22, 39,
    88–89, 228, 231; solution, 174; sound
    in, 230; windage, 130, 140, 217, 227
air conditioning, 218
aircraft, 21, 46, 180, 181, 182, 191
alcohol, 56, 57, 226, 234
Alder, Michael, 5, 6, 67, 145
alkaloids, 227
Allen, Bryan, 182, 191
aluminum: coating, 71, 106–7, 158–60;
    compounds, 91, 93, 94; metal, 87,
    140, 155–57, 177, 180, 222; powder,
    178
Amdahl, Gene, 30
amino acids, 116
Andrews, David, xi, 47, 58, 136–39,
    141–42, 151
anesthetics, 234
angina, 187, 188, 190, 191
anima, 233

animals: behavior, 14, 16–17, 36, 102,
    195, 227; exploitation, 223; mind
    of, 3–6, 8, 12, 66, 72, 80, 233, 235;
    organs of, 215, 223, 232, 235, 237
Anova (Daedalus invention), 78
anteaters, 8
anthropic principle, 242
apartments, 16, 59
aquarium tanks, 102, 164
Arcton (gas), 102
argon, 59, 102, 221
Arlen, Harold, 21, 55
armor, 119
arsenic, x, 235–37, 248
astronauts, 81, 114, 202, 226
Atkinson, Bruce, 84
atom bomb, 241
atoms, 93, 110, 196, 198, 241

Babbage, Charles, 62
babies, 21, 61, 80
Bacon, Francis, 46
Bakelite, 169, 206
ball bearings, 167
balloon: bursting, 229; cosmological
    analogy, 113; gases in, 101–2; hot air,
    xi, 108; steam, 61, 71, 106–8
bathtub temperature, 191, 247
battery, 215
BBC. See British Broadcasting Corporation
beach ball, 176, 177
Bean, W. B., 192
Beckman Company, 21
Bell Laboratories, 19, 26, 213

Hebb, Donald, 233
Herendeen, R. A., 213
heresy, 6, 33
Herken, Gregg, 13
Hoffnung, Gerard, 230
hollow tube, 139, 140
holography, 26, 65
honey spoon, 127–29
Honig, William, 202
hormones, 76, 78. *See also* estrogen;
  testosterone
Housman, A. E., 56
Howe, Elias, 9
Hoyle, Fred, 30, 65, 68, 120
Hughes, Ted, 7, 40
humidity, 219, 220
humor, 11, 33, 50, 51, 66, 207, 248
hydrogen: in astronomy, 120, 121, 241;
  in chemistry, 59, 87, 93, 142–44, 146,
  198; as lifting medium, 105
hypochondria, 185, 187, 188, 191, 192

IBM. *See* International Business
  Machines
ICI. *See* Imperial Chemistry Industries
igniter wick, 141, 149, 150, 155
Imperial Chemistry Industries (ICI), xi,
  2, 28, 95, 96, 246
Imperial College, 30, 54, 82, 113, 193, 204
incubation, 48, 50, 65
injector, 82, 84, 88, 89, 90
inquisitiveness, 14, 31, 246
insecticides, 227
insects, 195, 227, 228
interference, 109, 134, 135
internal combustion, 142, 215, 217
International Business Machines (IBM),
  29, 30, 42, 96
Internet, 12, 32, 33, 211
intuition: human, 60, 194, 243; large,
  82, 103, 130–31, 134, 175, 177, 247;
  tiny, 23, 70–73, 91, 106, 111, 227
inventions: methods, 82, 141, 161, 181,
  227; needed, 212, 216, 221, 223, 224;
  objects, 22, 28, 34, 42, 46, 140
Irving, David, xi

Jamison, Kay Redfield, 66
Janata, Jiri, 29, 30

Jeans, James, 121
Jennings, Paul, 208
Joule, James Prescott, 100
journals, scientific, x, xii, 24, 95, 202.
  See also papers, scientific; *individual
  periodicals*
Jung, Carl, 69
junk, 54, 131, 158, 173, 246

Kapitza, Peter, 28
Kekulé, Friedrich, 10, 11, 196
Kennedy, John F., 44
kerosene, 106
kettle, 102–4, 106–8, 170, 176
Keynes, John Maynard, 67
Kipling, Rudyard, 55, 67
kitchens, 1, 3, 7, 14, 41, 48, 132, 218
Koestler, Arthur, 11, 41
Krakatoa, 238
Kramp, Klaus, 86, 88

Laithwaite, Eric, 11
Land, Edwin, 161, 164
Langbein, Dieter, 89
language, 12, 44, 62, 201, 202, 206,
  207, 243
lasers, 68, 116, 196
leaks, 85, 87, 88, 90, 105, 159, 173, 218
learning: conditions, mechanisms of, 35,
  38, 51, 233; experience, 36, 40, 47,
  93, 166, 243; informal, 32, 42, 70,
  157, 169, 175, 230; teacher and
  media, 30, 76, 113, 135, 211
leaves, 155, 227, 231
lectures: mine, 122, 136, 142–43, 146–
  47; other people's, 10, 19, 23, 39, 89,
  136, 174
Ledingham, Kenneth, 236
letters, crank, 202, 203, 204
Levi, Primo, 44, 55, 204, 206, 222
ley lines, 111, 112
library, 32, 52, 176, 187, 202, 208, 230
light-emitting diode, 216
light, polarized, 160–62, 164
lightning, 230, 231
lights, 132, 146, 161, 229
liquefied petroleum gas (LPG), 148
literary styles, 201, 202, 204, 224
literature, 110, 202, 224